European Agriculture: Policy Issues and Options to 2000
An FAO Study

A shorter version of this study was originally issued for the Sixteenth FAO Regional Conference for Europe, Cracow, Poland, 23–6 August 1988, as document ERC/88/INF/4. Following the Conference, the study was updated, revised and enlarged to produce the present version, as indicated in the Acknowledgements. The study will also be published in French by Economica, Paris, as Alexandratos, N. (sous la direction de), *L'Agriculture européenne: Enjeux et options à l'horizon 2000, Étude de la FAO.*

European Agriculture: Policy Issues and Options to 2000

An FAO Study

Edited by

Nikos Alexandratos

Published by arrangement with the Food and Agriculture
Organization of the United Nations by Belhaven Press
(a division of Pinter Publishers) London 1990.

Food and Agriculture
Organization of the United Nations
Rome

Belhaven Press
A division of Pinter Publishers
London and New York

© 1990, Food and Agriculture Organization of the United Nations

First published in Great Britain in 1990 by
Belhaven Press (a division of Pinter Publishers),
25 Floral Street, London WC2E 9DS and P.O. Box 197, Irvington, New York

The designations 'developed' and 'developing' economies are intended for statistical convenience and do not necessarily express a judgement about the stage reached by a particular country or area in the development process.

The designations employed and the presentation of material in this publication do not imply the expression of any opinion whatsoever on the part of the Food and Agriculture Organization of the United Nations concerning the legal status of any country, territory, city or area or of its authorities, or concerning the delimitation of its frontiers or boundaries.

British Library Cataloguing in Publication Data

A CIP catalogue record for this book is available from the British Library

ISBN 1 85293 119 1

Library of Congress Cataloging-in-Publication Data

European agriculture : policy issues and options to 2000 : an FAO
    study / edited by Nikos Alexandratos.
        p.   cm.
    Includes bibliographical references.
    ISBN 1-85293-119-1
    1. Agriculture and state--Europe. I. Alexandratos, Nikos.
II. Food and Agriculture Organization of the United Nations.
HD1918.E97   1990
338.1'84--dc20   90-34582
                                                                        CIP

Typeset by Communitype Communications Limited
Printed by SRP Limited

CONTENTS

List of tables

List of figures

List of boxes

FOREWORD

By the Director-General of FAO

This study analyses the major options for agricultural policies in the countries of the European Region. An earlier and shorter version was presented as a background document to FAO's Sixteenth Regional Conference for Europe in August 1988 and was intended to enhance the understanding of the main issues facing European agriculture and to facilitate dialogue towards harmonized, multilateral action. Its coverage is extended to the countries of North America which participate in the Conference as members of the Economic Commission for Europe of the United Nations.

In Western Europe and North America agricultural support policies have raised production in excess of domestic and export demand; this in turn has led to serious imbalances in the major commodity markets and to trade disputes. Although the costs of implementing such policies have risen dramatically, some objectives such as farm income support have been only partially achieved. In some countries of the region the quest for policy reform is motivated by the need to improve agricultural productivity, to contain the costs of food subsidies and to enhance the supply of food, particularly as regards variety and quality. In all countries there is a growing awareness of the economic gains to be obtained from a more efficient use of resources.

These concerns have found expression in a series of initiatives for policy reform, at both national and international levels. At the international level there are a number of significant initiatives aimed principally at remedying the distortions in trade resulting from agricultural protectionism. These include the Ministerial Declaration of Punta del Este which launched the Uruguay Round of Multilateral Trade Negotiations and several proposals made at high political level on principles which can lead to greater liberalization and more discipline in agricultural trade. These principles were further elaborated in the agreement on agriculture reached at the Mid-Term Review of the Uruguay negotiations in Geneva in April 1989.

Major initiatives at the national level include the US Food Security Act of 1985 and the decision of the European Community on 'agricultural stabilizers'. The initiatives also include policies adopted in some centrally planned economies of Europe which relax the rigidities of central planning and encourage innovative forms of organization of production and marketing with emphasis on economic motivation, including, in some cases, the transition towards a system of market economy in agriculture.

A key issue underlying the various proposals is how to reconcile the conflicting objectives of improving market access and reducing or eliminating those subsidies and other measures which distort trade but without jeopardizing other domestic objectives such as food security, support of farm incomes, preservation of rural communities and the promotion of agriculture in disadvantaged regions. Significant differences exist between countries on the relative weight assigned to domestic policy objectives, reflecting differences in agricultural resource endowments and socio-economic conditions. The fact is that there are large numbers of farmers whose disadvantaged position calls for some measure of support and assistance for adjustment. Thus, the issue in the international context is not whether support may be provided but rather how it should be applied with minimum distortions in the world markets for agricultural products.

This study surveys and analyses the policy experiences and major options available for this purpose. These range from price policies through different measures for supply management to the provision of direct income support. Environmental factors are especially singled out as important elements to be considered in the design and implementation of agricultural policies. The importance of international cooperation in integrating environmental factors into agricultural policy is emphasized.

Developments in agricultural policy reform in the region must also be seen in the global context of food and agriculture. That is why the study examines the policy issues of concern to the region in the context of the findings of the revised study *World Agriculture: Toward 2000*, which was discussed at the twenty-fourth FAO Conference in November 1987. Notwithstanding the recent tightening of supplies in world food markets, excess production capacity in the region continues to coexist with severe inadequacies in levels of consumption and nutrition in the developing world.

The global study concludes that, in the foreseeable future, the demand of the developing countries for food exports from Europe and North America is unlikely to be the dynamic element in world agricultural trade that it was in the past, particularly in the 1970s. This is mainly the consequence of poor prospects for the growth of incomes and foreign exchange earnings of the majority of the developing countries. Activating this source of latent demand for the agricultural exports of Europe and North America through more liberal policies on trade (including agricultural trade), debt and resource flows could make a significant contribution in stimulating agricultural trade. More generally, policy reforms in the food and agriculture sector must, on the one hand, be integrated with the reforms for the overall economy in the different countries of the region and, on the other hand, foster a more balanced economic and social development between the industrialized and developing countries.

In conclusion, I must underline that in many respects the needed adjustments in the principal countries of the region are the result of success, not failure, of past policies in raising production and productivity. Further increases in productivity, including the longer-term gains from the application of biotechnology, present policy makers with a challenge as well as an opportunity. The challenge is to resist the temptation to accommodate the past gains by further protected import substitution and subsidized exports at the expense of low-cost producers. The opportunity is to take advantage of productivity gains to promote farm structures that are economically viable given an enhanced role for market forces in achieving the right balance.

Improved cooperation in agriculture would help harmonize policies and facilitate transition towards national agricultures which are economically viable and much less in need of strong support. It would also enable all countries, particularly those endeavouring to increase their production growth rates, to benefit from the wider diffusion of new technologies and the experiences of other countries. In this regard FAO has a useful role to play in support of European cooperation. It is my hope that this study will provide a proper perspective for cooperation in food and agriculture, at a moment when the significant policy reforms currently in progress call for enhanced cooperation in Europe. The Organization stands ready to continue and strengthen its cooperation with Governments, the UN Economic Commission for Europe and other institutions in the countries of the region.

Edouard Saouma
Director-General of the FAO

Acknowledgements

A shorter version of this study was originally prepared in 1988 by a team led by N. Alexandratos and comprising Jelle Bruinsma (FAO), Alan Matthews (Trinity College, Dublin) and Trevor Young (University of Manchester). Milan Trkulja on behalf of the Director-General of FAO supervised and guided the whole effort. The study was revised and enlarged in 1989 mainly by the editor, under the supervision of B.P. Dutia, Assistant Director-General, Economic and Social Policy Department, FAO. The following persons also contributed to the study at different stages of the work and in various ways: FAO staff: H. Hjort, R. Perkins, D. Norse, J. Greenfield, E. Littmann, J. Budavari, M. Brzoska. Non-FAO: S. Magiera (USDA). Many others both within and outside FAO contributed to the work of the team or provided useful comments and suggestions for improvements. Maria Grazia Ottaviani and Isabel Reyes Hormazábal provided efficient computational and statistical assistance. Monica Brand Roberti assisted by Patrizia Mascianà provided secretarial assistance to the team.

The study benefited from comments received during the review of an early draft by a group of external experts (listed below) at a meeting held on 28–9 March 1988. The present version of the Study does not necessarily reflect all comments and suggestions made during this meeting.

K.C. Clayton, Office of the US Trade Representative, Washington DC, USA.

C. Csaki, Karl Marx University of Economic Sciences, Budapest, Hungary.

K.H. Lagerfelt, Ministry of Agriculture, Stockholm, Sweden.

A. Papasolomontos, formerly Ministry of Agriculture, Nicosia, Cyprus.

H.W. Popp, Department of Agriculture, Berne, Switzerland.

P. Scandizzo, University of Rome (Tor Vergata), Italy.

A. Simantov, Centre International de Hautes Etudes Agronomiques Méditerranéennes, Paris, France.

M. Tracy, Belgium.

A. Wos, Institute of Agricultural and Food Economics, Warsaw, Poland.

1 Introduction and overview

Background

The present report is a sequel to the FAO study *World Agriculture: Toward 2000* (Alexandratos 1988). The above study examined the prospects and policy issues for world food and agriculture based on an analysis which covered virtually all the countries of the world, but with more attention paid to the developing countries. The position of the European region in the world food and agriculture system was defined with a broad brush and analysed principally in terms of policy issues relating to trade and adjustment. The present study analyses these issues in somewhat more detail with more attention to the domestic policy concerns which are often at the root of issues relating to trade and adjustment. The objective is to raise issues and discuss options, not to provide answers, much less to prejudge the possible outcomes of current efforts at policy reform.

Policy issues and options are examined in the context of a quantitative framework of projections of production, consumption and trade of agricultural products to the year 2000. The nature and relative importance of the different issues and the analysis of possible policy responses depends so closely on the probable evolution of these key variables of agriculture (e.g. the extent of demand slowdown, the rate of growth of agricultural productivity, the projected size of the net import requirements of the rest of the world) that it would be very difficult to discuss issues and policies without reference to such projections, including the issue whether in Western Europe a mere slowdown in the growth rate of production is required or whether absolute declines in output are necessary.

This study covers all the countries which are members of the FAO European region, as well as Canada, the German DR, the USA and the USSR, members of the UN Economic Commission for Europe. The 33 countries of the study (sometimes referred to collectively as 'region') as well as the groups used in the text and tables are shown in the Appendix. The study examines the policy issues of the crops and livestock sectors only. Time limitations did not permit analysis of issues relating to forestry and fisheries, although these two sectors are a very important component of total agriculture in the wider sense, particularly in some countries, e.g. Sweden, Norway, Iceland, Portugal. Issues relating to European forestry to the year 2000 were analysed in a study prepared in 1986 (UN/ECE/FAO 1986).

The discussion of policy issues and options in this study must be seen in the context of current national and international efforts towards agricultural policy reform. In the last few years there has been increasing dissatisfaction with the workings of national agricultural policies. In some major countries or country groups of the region such policies were overachieving some of their objectives, e.g. production growth, while largely missing others in varying degrees, e.g. farm income objectives. As a result, serious imbalances developed in the markets for the main commodities, the risk of trade conflicts increased, with potential spillover effects from agriculture to other sectors, and the costs of agricultural policies (to consumers, the budget and the economy as a whole) rose spectacularly. In other countries, particularly some Centrally Planned Economies (CPEs), dissatisfaction with the workings of the agricultural policies centred on the low productivity achieved in the resources employed in agriculture, and underachievement of production growth targets and inadequacy of food supplies, particularly as regards variety and quality, which was partly attributable to the poor performance of the downstream sectors of agriculture (transport, storage, processing and marketing). Additionally, the budgetary costs of agricultural and food policies rose spectacularly, particularly the costs of food subsidies in some CPEs.

The increasing awareness that such policies were not viable and had to be modified is reflected in a number of major policy statements and/or initiatives, which provide the background for the design and implementation of policy reform. These include:

1 In launching the Uruguay Round of Multilateral Trade Negotiations in September 1986 the GATT contracting parties agreed to give special attention to agriculture. Specifically, the negotiations 'shall aim to achieve greater liberalization of trade in agriculture and bring all measures affecting import access and export competition under strengthened and more operationally effective GATT rules and disciplines'. In particular, the Ministerial Declaration launching the new round refers to improving market access through such measures as reduction of import barriers; increasing discipline on the use of all direct and indirect subsidies and other measures affecting agricultural trade, including the phased reduction of their negative effects and dealing with their causes; and minimizing the adverse effects that sanitary and phytosanitary regulations and barriers can have on trade in agriculture. Following the Mid-Term GATT Review held at Montreal in December 1988 and at Geneva in April 1989 the contracting parties agreed to the long-term objective of policy reform to 'establish a fair and market-oriented agricultural trading system' and 'to provide for substantial progressive reductions in agricultural support and protection sustained over an agreed period of time, resulting in correcting and preventing restrictions and distortions in world agricultural markets'.

2 Similar principles are enunciated in the OECD Ministerial Meeting Communiqué of 13 May 1987 which states *inter alia* that

'the long-term objective is to allow market signals to influence by way of a progressive and concerted reduction of agricultural support, as well as by all other appropriate means, the orientation of agricultural production; this will bring about a better allocation of resources which will benefit consumers and the economy in general. In pursuing the long-term objective of agricultural reform, consideration may be given to social and other concerns, such as food security, environment protection or overall employment, which are not purely economic. The progressive correction of policies to achieve the long-term objective will require time: it is all the more necessary that this correction be started without delay.'

3 Reforms in the Common Agricultural Policy of the European Community (EC) contained in the conclusions of the European Council of 11–13 February 1988 aim at bringing agriculture into the mainstream of efforts to make a success of the Single European Act. They would enable agriculture to contribute to the more efficient use of resources and to bringing about required adjustments in the EC economies. To this effect, policy measures (agricultural stabilizers) were adopted to control production and agricultural expenditures. If successfully implemented they would contribute to improved adjustment of supply to demand by enabling the market to play a greater role.

4 The 1985 Food Security Act of the USA must also be listed among the major policy initiatives of global significance undertaken in recent years. The Act aimed at moving the world's leading cereal exporting country to a more competitive position in world agricultural markets. It reduced loan rates (and thus the *de facto* floor prices in many commodity markets) and gave the Secretary of Agriculture added flexibility to change programme parameters as market conditions warranted. The Act also specifies acreage reduction programmes when stocks exceed certain levels and it attempts to control erosion and prevent farm production in highly erodible lands through a Conservation Reserve Program. The Act expires in 1990. The issues raised in the preparation of the new farm legislation in the USA are currently being debated in the US Congress (for discussion see Lipton 1989).

5 In the Centrally Planned Economies (CPEs)[1] the current efforts towards policy reform emphasize relaxation of central planning and the associated bureaucratic control, devolution of more management and financial responsibilities to individual enterprises and resort to innovative forms of organization of production and marketing (e.g. the family and brigade contract systems, the leasing of land, and more generally the promotion of a symbiotic relationship between private and socialized farming) to stimulate economic motivation towards increasing production and improving efficiency and quality. Some of these measures are akin to those already applied in the past in some CPE countries, most notably in Hungary (for an analysis and evaluation of the Hungarian experience in the mid-1980s, see Kočnai 1986). In parallel, some countries are moving

---

[1] The traditional term CPEs is used to refer collectively to the countries of Eastern Europe and the USSR for convenience of presentation only. Although the term conveys an idea of homogeneity of economic systems it should be noted that differences do exist among countries in this group. Differences have tended to increase in recent years with the reforms that are under way, which, for some countries, involve radical system change and transition towards systems of market economy.

further ahead towards the market economy system.[1] Reforms received a new impetus in the context of the more general efforts at policy reforms aimed at accelerating the rate of economic and social development in the USSR, particularly after Mr Gorbachev's policy statement at the Congress of the collective farms in March 1988 and the mid-March 1989 Plenum of the Central Committee of the Communist Party of the USSR devoted to food and agriculture. Agriculture has been singled out as a priority sector in applying policy changes both because of the urgency of the need to improve food supplies but also because it is expected that positive results would be obtained more quickly compared with other sectors.[2]

The last two years (1988 and 1989) have witnessed considerable tightening of the demand/supply situation, mostly on account of the North American drought of 1988, whose effects were superimposed on the largely policy-induced output declines of 1987. The resulting declines in cereal stocks and sharp rises in international prices coincided with the heightened awareness and concerns for the environment and the associated perceptions of possible climatic change. Some interpreted this situation to be more than just another temporary deviation from the long-term path of increasing global supplies per caput accompanied by gently declining real food prices. Questions have been raised as to whether the world is set on a path of increasingly tight supplies. The underlying concerns relate to the capacity of the world's land and water resources for continuation of agricultural growth not only at rates above those of world population but also at rates adequate for meeting growth of effective demand at non-increasing prices; perceptions that these resources are being degraded by over-exploitation; and the extent to which progress in agricultural technology tends to level off.

Official reactions to such concerns tend to consider the present situation of tight supplies as a temporary phenomenon, implying that there is no real reason for letting up on policy reform to correct the support policies that were responsible for over-production in the recent past. The Council of the OECD which met at ministerial level on 31 May and 1 June 1989 considered that

the role of market signals in orienting agricultural production remains insufficient almost everywhere ... it is therefore more than ever necessary that the process of agricultural reform be pursued vigorously, in conformity with the principles defined by Ministers in 1987 and 1988, and taking advantage of the present strength of markets. (OECD 1989c)

[1]   'We are moving with determination towards a market economy ... In agriculture ... we are allowing free play to the working of market laws' (Statement to the 25th Session of the FAO Conference, November 1989, by the Minister of Agriculture, Forestry and the Food Economy of Poland — FAO document C 89/PV/9).

[2]   'I am convinced that the most effective form of organizing production on the basis of full cost accounting will take root quickest in the agro-industrial complex ... rural folk are enterprising and resourceful. All this makes for greater mobility and flexibility when applying cost-accounting, self-sufficiency and self-financing' (Gorbachev 1988: 96). (See, however, Ch. 8.)

Similar views are echoed in the latest report of the Commission of the European Communities on the agricultural situation in the Community in 1988:

It is true that, as regards world markets, there has, since early 1988, been a definite improvement in the situation for certain key products such as cereals, oilseeds and milk and milk products. The tendency for supply to run well ahead of demand for these products is not nearly as marked as it was. But it must also be recognized that this improvement is largely incidental or accidental, rather than structural, in character. It does indeed, for the time being, facilitate market management, but it does not really get to the root of the problem. Therefore, the reform policy must be pursued and consolidated. (CEC 1989a: 26–7)

Finally, a recent USDA study concludes that 'On balance, it appears that in the next two to three decades the long-term phenomena that are causing concern do not pose an insurmountable obstacle to the world food system. In this time frame, food crises will likely continue to be localized and to be the consequence of policy-induced inflexibilities in the world trading system and lack of effective demand, rather than of failure of supply' (USDA 1989). It is also possible that biotechnology applications expected from the mid-1990s onwards will further strengthen trends towards increased productivity, at least in Western Europe (CEC 1989d). These perceptions lead to the conclusion that the recent relatively tight market situation should be followed by a period when global agricultural production will rebound and the increase in output is more likely to be constrained by insufficient growth in effective demand.

Different agricultural policy profiles

The countries of the study span a wide range of agricultural situations, levels of development, socio-economic systems and experiences in agricultural policies. Although all of them pursue a set of similar objectives in agricultural policy (e.g. food security, stable consumer prices, farm income support), in practice there are wide differences in the interpretation and relative weights attached to each objective and in the instruments used to achieve them. Even such a mundane objective as efficiency of resource use in agricultural production and the economy as a whole can mean very different things to different policy-makers. These differences can be ultimately traced to the way things agricultural (inputs, outputs, environmental aspects) are valued in the policy-makers' calculus. Some countries are comfortable with the valuation given to resources and outputs by the market while others will not accept such valuation because they consider that farming has significant 'externalities' not reflected in the market prices. These are variously perceived as resulting from failures of the market to attach appropriate values to food security, preservation of traditional rural life styles, balanced regional development, environmental and amenity aspects, etc. The policy-maker, therefore, steps in to account for these externalities by altering some basic parameters of the system.

One can hardly quarrel with this idyllic view of the policy-makers' functions which, however, is an over-simplification of reality. Rarely does society express its preferences in a neat rational manner and equally rarely do policy-makers have the required predictive power or find it practicable to translate objectives into a consistent set of policy instruments. As a result agricultural policies, as indeed any policies, can exhibit some serious shortcomings in the sense that they may have a net effect opposite to the one pursued, may be achieving one professed objective at the expense of an equally preferred one, or they may be inefficient, i.e. unnecessarily costly for what they achieve.

In addition, the increased integration of agriculture into the national and the world economies has meant that many traditional objectives of agricultural policy are influenced, more than in the past, by events and policies outside agriculture, e.g. those that affect the money supply and the interest and exchange rates. For example, farm income problems in the first half of the 1980s were partly caused by high interest rates that farmers had to pay for their loans. Similarly, the wide movements in the exchange rates over the same period have been a key determinant of the comparative growth rates of agricultural exports and of public expenditure on agriculture in the USA and the EC. In these conditions it becomes increasingly difficult to achieve the objectives by means of agricultural policy instruments alone, or even predominantly, even when the objectives are strictly agricultural. The picture can be further complicated when agricultural policy instruments (e.g. support prices) are used to achieve objectives which are not strictly agricultural, for example, regional development and income redistribution.

In recent years there has been heightened awareness of the need for policy reform as evidence has been increasing that existing policies often fail to achieve fully their objectives, e.g. restoration of market balance, support of farm incomes and containment of costs in the major Western countries, or increased production and improved supplies and quality of food products in the CPEs. In the major Western countries, particularly the large exporters, added impetus for policy reform has come from the need to reduce the trade distortions and potential for conflict resulting from failure to adjust their agricultures in a timely and coordinated fashion to the slower growth of markets of the 1980s.

Each country has its own characteristics and specificities and it is not possible to speak of policy issues and options for the countries of the study as a whole. At the same time there are enough similarities among particular countries to provide a broad basis for taxonomy, indicated in the following paragraphs.

The *Centrally Planned Economies* (CPEs) have traditionally attached high priority to the achievement of self-sufficiency and the maintenance of low food prices to the consumers, though this latter objective has been relaxed in recent years in many countries. It has also been characteristic of many CPEs that availability of food at affordable prices, particularly of meat, is a key variable influencing popular perceptions as to the economic well-being of the country and, of course, a key political variable in all countries. The CPEs are, however, far from being homogeneous as regards agriculture, for two reasons: first, the organization of the sector ranges from mainly private (Poland) to mainly cooperative (Hungary) to various combinations of cooperative and state farming (other countries, including, in some of them, a role for the subsidiary private sector), and second, policy experiences range from main reliance on central planning to nearly complete freedom as regards production and marketing decisions. Indeed some CPEs (e.g. Hungary) have experimented for a number of years with policies akin to those proposed now in other countries (e.g. the USSR) in the context of policy reform. The result has often been a symbiosis between private and socialist farming systems or, perhaps more correctly, between market-signal based and central-plan based decision-making processes. More recently, some countries have made further moves towards a market-economy system. It must be noted, however, that reforms currently underway in many CPEs mean that policies are under continuous evolution and often involve radical, rather than fine-tuning, changes. It is not possible, therefore, to be definitive as to how the policy scene for food and agriculture may crystallize in the end.

The impetus for agricultural policy reform in the CPEs is part and parcel of the more general thrust towards policy reform in the quest for improved performance of their economies. Reforms in a number of countries aim at relaxing some of the rigidities of central planning and at allowing more scope for decentralized decision-making — including on the basis of economic incentives and market signals — while in other countries they aim at transition to full market mechanisms. Policy reform in agriculture must, therefore, be seen in the wider context of general reform. Given the increasing links of modern agriculture with the rest of the economy, the success of agricultural policy reform may well depend on the extent to which it is coordinated with reforms in other sectors and in the economy as a whole. This raises the question whether in a set-up of central planning, the introduction of market elements selectively in one sector or in one market only can be a success, e.g. reforms in product but not factor and asset markets, or decentralization of decision-making and shift of enterprises to full financial accountability but with varying degrees of liberalization of production, pricing and marketing decisions. This is a wider question which cannot be explored in this study, but it may be of crucial importance in determining the chances of obtaining durable and continuous, rather than one-off, benefits from policy reform in agriculture. There is a risk that policy interventions in food and agriculture would remain haphazard if they are not part of a wider coherent strategy of transition to a restructured economic system. Further, because radical rather than marginal changes are often involved, knowledge required for formulating such a strategy is still inadequate, in particular as regards the sequencing and pace of implementation of the reforms contemplated.

In the majority of the countries of *North-western Europe and North America* the impetus for policy reform is rooted in the continued strong growth in agricultural production in the face of significant slow-down in the rate of expansion of domestic and export markets for temperate zone agricultural products and the virtual exhaustion of the scope for further import substitution. These two latter sources of market expansion were instrumental in the past in maintaining agricultural growth above that of the continuously decelerating domestic demand. In the 1980s and up to quite recently, publicly financed stock accumulation and disposal at give-away prices helped sustain the growth of production above equilibrium levels, but this is only a temporary palliative. In the end, countries are being forced by circumstances to cut their agricultural growth rates. At issue is the extent to which farm income and similar cherished objectives of agricultural policy can be achieved and at the same time market imbalances redressed, trade distortions reduced, budget costs contained, and economy-wide efficiency in resource use improved.

These common issues notwithstanding, this group of countries is far from homogeneous. Differences in policy attitudes abound. The main split is between those who hold the position that agriculture is just another industry in which policy interventions should not distort markets, e.g. farm income support should be by means which as far as possible do not affect production decisions; and those who consider that the high value placed by society on the significant 'externalities' of agriculture justify some (and sometimes substantial) deviation from market solutions: they would at most concede that the effects of their policies should be limited to their domestic economies and should not spill over into the international markets in the form of subsidized exports or 'excessively high' rates of self-sufficiency, although this is not equivalent to saying that such policies may not distort international markets. These differences reflect both different objective economic conditions of agriculture (e.g. resource endowments and farm size, regional imbalances, export orientation, etc.) as well as different historical experiences and cultural backgrounds leading societies to attach different values to food self-sufficiency and preservation of farming. Differences notwithstanding, the unifying theme of agricultural policy-making in the great majority of the countries in this group is the small relative economic size of the sector in what are essentially high-income mature industrial economies: they can afford to attach high relative weights to the non-strictly economic objectives of agricultural policy. In these circumstances, purely economic arguments for agricultural policy reform have only limited impact except, perhaps, when efficiency losses of agricultural support policies hurt other vital economic sectors, including in the context of international trade.

The situation is different for most countries in *Southern Europe* in which the objectives of agricultural policy and the context within which they are pursued reflect their lower per caput incomes and the associated greater weight of agriculture in their economies as well as their different ecology and product composition of agriculture. Lower incomes and greater shares of agriculture in the economy and employment mean that most countries of Southern Europe and, at the other geographical extreme, also Ireland, can afford much less than the previous group of countries agricultural policies which are costly to the rest of the economy in terms of income transfers and/or overall efficiency losses. Some countries in this group may simply deviate from that most European of all agricultural policy characteristics, namely the direction of intersectorial resource transfers to, rather than out of, agriculture. The different product composition of their production and trade also puts them apart. It means that they are interested in access to low-priced imports of cereals and livestock products and high prices for their typical export products (citrus, fruit and vegetables, wine, tobacco), some of which (e.g. table wine, tobacco) also suffer from structural surpluses.

Related to the above and contrary to the previous group of countries in which competition comes predominantly from within the group, the Southern countries compete to a much larger extent with produce from outside the European region, much of it from the developing countries. For them, policies have to pay attention to conditions in these competing countries as well as to the policies of their European partners *vis-à-vis* these latter countries. In discussing the relations of the European region with the developing countries these differences must be kept in mind.

The importance of *national* agricultural and overall economic characteristics in agricultural policy-making has been considerably lessened in the EC countries by the existence of the *Common Agricultural Policy* (CAP). The countries of the EC-12 span almost the entire range of ecological and economic conditions encountered in Europe. The sheer variety of farming within the EC-12 and the many functions of agriculture recognized in the different countries has meant that the design and operation of the CAP has been fraught with difficulties. The supranational character of the CAP means that the agricultures of the individual countries are affected by policy decisions taken on the basis of criteria reflecting European-wide interests (including the wider political and economic interests of European unification) and relative bargaining strengths of the different member countries. As a result of the CAP, agricultural policies and agricultural situations in the EC-member countries are different than what they would have been in the absence of the CAP. Generally speaking, the net effect has been that the overall level of protection of European agriculture as well as total farm output are above those that would have prevailed without the CAP. This has been the result of policies aimed at implementing the basic CAP principle of community preference which favoured import substitution. The financing of the CAP by community funds, and the emphasis on farm income support and reduction of regional income disparities further reinforced the policies leading to increased production.

The successive enlargements of the EC with the addition of predominantly lower income and more agriculture-dependent countries and regions has meant that first, regional income disparities within the EC increased (Padoa-Schioppa 1987) and second, the proportion of the population dependent on agriculture for their incomes also increased. These developments occurred at the time when the scope for increasing farm incomes by means of further output growth became progressively more limited (both because of market limitations and because policies which stimulated output increased the incomes of the large rather than the small farmers), budgetary problems became more severe and the external environment became very hostile to further EC increases of subsidized exports and import substitution. Under the circumstances, the emphasis of the CAP in the future may be expected to be much less on output-increasing policies compared with the past and more on structural policies and supply management measures.

Key Parameters of the demand-supply balance to 2000

Continuing slow-down in the growth of domestic demand

For the region as a whole the growth of *food demand* for farm products would continue to slowdown, to approximately 0.9 percent p.a. over the period to 2000 from 1.3 percent p.a. in the 15 years to 1985. One half of this slowdown is due to lower population growth and the balance to the limited scope for further increases in per caput consumption, though these factors are much less restrictive in some countries compared with the average. Thus, food demand may grow at over 2 percent p.a. in some Mediterranean countries compared with near zero growth in some countries of Western Europe. Likewise, there is significant scope in some CPE countries for increases in per caput consumption of particular products, e.g. meat, fruit, including imported fruit, and tropical beverages. The above growth rates refer to the primary product equivalent (farm value) of food consumption which is normally less than one half of the retail value of food purchases. Actual consumer expenditure on food would grow faster than indicated above reflecting the ever increasing margins of processing and distribution.

The other major component of domestic outlets for the gross output of agriculture, *animal feed*, would also exhibit slower growth compared with the past. The factors behind this deceleration include slower growth of the livestock sector (following the slowdown in food demand growth), continued technical progress and related productivity increases in animal production and possible shifts to more grass and roughages feeding following land released from crop production. Overall, therefore, *aggregate domestic demand* in the region for the gross output of the sector may grow at a rate not exceeding 1.0 percent p.a., continuing the trend towards slowdown, from 2.6 percent p.a. in the 1960s to 1.3 percent p.a. in the 15 years to 1985.

Less scope for export growth and import substitution

For the western countries of this study, particularly North America and the EC, the rapid expansion of food exports to the developing countries and the CPEs in the 1970s and the displacement of imports, particularly in the EC, were instrumental in maintaining the growth rate of agricultural production well above that of domestic demand (see Figs 4.1 and 4.2 for changes in the net trade positions and the self-sufficiency rates). These factors weakened considerably after the early 1980s as the room for import substitution was virtually exhausted and exports suffered from the slowdown in economic growth and foreign exchange shortages in both the developing countries and the CPEs as well as from reduced import requirements following improved production performance in some major importing countries.

These factors depressing the markets for food exports from the Developed Market Economies (DMEs) are expected to persist over the period to the year 2000. It is assumed that the CPEs would continue to pursue their objective of increasing self-sufficiency, including for reasons other than purely economic ones. The growth rate of their demand would continue to be strong, compared with that of the DMEs, though somewhat below that of the historical period, given the high consumption level achieved in some countries, the emphasis on quality rather than quantity and the possibility of more restrictive policies concerning food subsidies. Policy reform under way could lead to somewhat higher production growth compared with the past, particularly in the USSR. All these factors could lead to their net food import requirements not growing above current levels, or actually falling, which would be a departure from the past trends of rising net imports. Other studies project a wide range of possible net cereal deficits of the CPEs, ranging from under 20 million tonnes to some 70 million tonnes (see Ch. 4) compared with some 35 million tonnes in the average of the last two years 1987–8.

In the developing countries net import requirements should continue to grow but at a much slower pace than during the 1970s. Slow economic growth and scarcities of foreign exchange, greatly aggravated by the debt problem, would continue to depress the prospects for food imports in many developing countries, notwithstanding the persistence of very inadequate levels of nutrition. No attempt is made here to incorporate into the analysis the possible occurrence of events which could result in windfall gains in the incomes and foreign exchange earnings of the developing countries similar to those that accompanied the oil sector developments in the 1970s. Should such events occur, however, they would most likely raise the developing countries' net food import requirements to levels well above those underlying this analysis (110 million tonnes of cereals by 2000, Table 4.6). Fluctuations in output due to weather variability are also likely to cause import demand in some years to be greatly in excess of or below the trend. Other studies have higher projected net import requirements (Box 4.1). The possibility that the food import needs of the developing countries could be significantly above those used in the present analysis has certain implications for the policies to be used in slowing down the growth of production in the DMEs as a whole. In particular, the changes brought about by these policies must be reversible (e.g. land set-aside programmes) rather than permanent (e.g. destination of surplus farm land to permanent non-agricultural uses).

In general, it is important to guard against the outlook for the longer term future being excessively influenced by the mood prevailing at the moment of writing. There is a risk that the current pessimism as regards the possibility for resumption of economic growth in the developing countries and optimism as regards the continued progress in agricultural technology may unduly influence the perceptions concerning the growth of effective global demand (inadequate) and supply (excessive). The excessive optimism of the 1970s as regards global demand, and pessimism as regards resource constraints to food production, provided an impetus to the adoption of expansionary policies and the creation of agricultural overcapacity in some countries of the region. Additionally, account must be taken of the risk that adjustment policies aiming at redressing market imbalances may — particularly when taken without coordination among major producers — overshoot their targets and lead to global shortages, especially when coinciding with unforeseen events such as weather-induced declines. The

1988 North American drought, whose effects were superimposed on the policy-induced reduction in production of the year before, is an appropriate reminder.

Production growth not to exceed 1 percent p.a.

The preceding discussion indicated that for the region as a whole the growth of aggregate production compatible with approximate balance in world markets should be around 1 percent p.a. from the mid-1980s to 2000, but higher if measured from the lower production levels of the last two years (1987, 1988) which resulted from policy changes and the North American drought. Some acceleration of agricultural growth would seem to be in order and likely to occur in the CPE region as a whole, while cutbacks would be required for the western countries as a whole, including the non-study DMEs, Japan and Oceania, compared with the longer term historical period, particularly the 1970s. It is important to recognize that what is required is a slowdown in production growth, and not its cessation or negative growth rates, over the medium to long term. Already the 1980s have witnessed a considerable slowdown in the growth rate of Western Europe and North America even before the drastic decline of 1988 due to the drought, to 1.0 percent p.a. in 1980–7 compared to 2.1 percent p.a. in 1970–80. How the total extent of adjustment may be apportioned between Western Europe, North America and the other countries is and will continue to be a key policy issue of international agriculture. The last few years have witnessed efforts to slow down the growth of production: small and gradual changes in Europe, e.g. in the dairy sector of the EC, bigger and sudden ones in the USA, e.g. cereals in both 1983 and 1987. The recent EC decisions on 'agricultural stabilizers' is a further step in this direction. There is certainly great need for policy coordination aimed at minimizing the social and economic costs of adjustment, with due recognition of the legitimate development and trade requirements of other countries, both developed and developing ones.

It is common to speak of the 'costs' involved in the transition to a lower growth agriculture. From a strictly economic standpoint, however, gains rather than costs would result from a lower growth regime if this is achieved by means of a lowering in the real rate of protection of agriculture and allowing more scope for market mechanisms to operate. The costs of agricultural policy to taxpayers and consumers (through higher prices) amounted in 1984–6 to some 80 billion ECU a year in both the EC-10 and the USA (OECD 1988). Only part of these costs are translated into benefits to the farmers while the balance is a net loss to the economy. At the macro-economic level, reduction of protection would relocate investment resources to higher productivity sectors and reduce the tax burden, thus increasing the overall income and employment. A number of studies indicate that the resulting gains can be significant although such estimates are inevitably surrounded by controversy.

Issues and possible policy responses

Shared responsibility for adjustment

The projections to 2000 bring out the extent to which agricultural growth in the Developed Market Economies must be reduced to a more sustainable rate. It is being increasingly realized that long-term viability of national agricultures depends on the achievement of better equilibrium between demand and supply, although opinions differ widely as to the role to be played by market forces in this task. It is important to emphasize at the outset that orderly transition to a slower growing agricultural sector should be considered a shared responsibility of all the countries which have in place measures sustaining their production at above equilibrium levels, not only those which subsidize exports. Moreover, multilateral coordinated action would apportion the burden of adjustment more equitably among countries and hence would be politically more acceptable. Yet the disposition of many countries towards such approaches has often been lukewarm. This attitude bears witness to the deeply ingrained feeling that agricultural policy-making is a predominantly internal question having to do with social structures, welfare and food security. As such, it has been viewed until recently as much less subject to international negotiations than policies in other sectors. Yet these very policies (indeed their very success in raising production) have often resulted in agricultural trends which make multilateral coordination all the more imperative. For example, rising agricultural productivity, to a large extent fostered by these policies, has turned importing countries into exporters, increased the visible components of total costs and heightened the potential for trade conflict.

Any country attempting to cope with these problems alone would need to bear a disproportionate share of the cost of adjustment: hence, the recently increased readiness of many countries to consider policy reform in a multilateral context. To judge from developments to date, only a few countries seem prepared to go all the way in subjecting their agricultural policies to international negotiation, although many favour coordinated policy reform that would enable the market to play a greater role. Some countries, however, are prepared to consider coordinated action aimed principally at correcting the excesses of agricultural trends, notably those which cause visible costs to increase beyond what is considered acceptable and those which heighten trade conflicts. Wider and largely concealed costs, notably those borne by the consumer and those implied by reduced overall efficiency in resource use appear to be less important as factors in the quest for policy reform, at least in some of the more prosperous countries of the region. Thus, these countries seem to favour agricultural policies affecting the great bulk of production to continue to be determined by the more traditionally 'European' motives: self-sufficiency, preservation of farming as a protected economic activity, maintenance of farm incomes and the pursuit of regional development by means of, among other things, agricultural support policies. Policy reform aimed at more efficient use of resources in both agriculture and the economy as a whole seems to be of comparatively higher priority in the CPEs. At the same time, however, many of these countries place even higher priority on food security and self-sufficiency. This often translates into reluctance to resort to trade as a means of increasing efficiency in resource use, although they often emphasize agricultural exports as a means of increasing foreign exchange earnings and refer to international market prices as one of the elements to be taken into account in the pursuit of domestic price reform. To the extent that policy reforms currently underway lead to greater integration of the CPEs into the world economy, their agriculture would also benefit and this would contribute to changing the context within which the objective of improving self-sufficiency will be evaluated.

Policies to stimulate domestic demand hold little promise

The prospects for increasing domestic demand (over and above the growth in demand projected in this study) through deliberate policies are not bright in most countries. Any potential for increasing food consumption through promotional expenditure or fiscal subsidies will be counteracted by nutrition policy considerations indicating the desirability of a further fall in food consumption per caput and in the consumption of red meat, dairy products and sugar particularly. The potential for significant increases in demand for some products, e.g. meat and fruit, particularly for the better quality products, exists in many CPEs and some other countries. Under favourable conditions (availability of supplies, incomes, prices) more of these products would be consumed than projected here, although the net effect on aggregate demand would probably be minor, particularly in the CPEs, because the increased consumption would be for substitutes for other, or inferior quality, products. More liberal food import policies in the countries whose policies constrain access to imported preferred foods (e.g. in the CPEs) may have a net demand increasing effect. Import displacement policies hold out little prospect of increasing demand for the region as a whole, although they would affect the distribution of the burden of adjusting to overcapacity. The development of non-food uses as a means of absorbing comparatively significant quantities of agricultural products, while technically feasible, is likely to remain uncompetitive for a long time. The major uncertainty on the demand side concerns the likely development of export markets, particularly in the developing countries, and to some extent the rate at which improvements in feed-use efficiency will be achieved in the livestock sector of the USSR. Faster and more equitable growth in the developing countries would increase the market for agricultural exports from the study region. The policies pursued by the study region countries, particularly with respect to trade, aid and debt problems, may have as much significance for the resolution of agricultural market imbalances as specifically agricultural policy measures.

Policies to control production growth in the market economies more likely to redress imbalances

Reduction of the growth rate of agricultural supply (after the recovery of production from the 1988 drought) to a more sustainable rate as determined by market opportunities will continue to be among the main objectives of policy in many countries, although the opposite would be the case in other countries, particularly the CPEs. Reduction in the growth rate can only be achieved by two types of adaptation: either deliberate efforts are made to *slow the rate of factor productivity improvement*, or total *factor input is reduced*. Concerning the first objective, account must be taken of the prospect that technological advances will continue to increase the productive potential of the resources employed in agricultural production, and possibly at an increasing rate. Public expenditure on agricultural research continues to increase in real terms, and what evidence there is suggests that expenditure by private firms is increasing even more rapidly. The commercialization and rate of adoption of the stream of innovations emerging from research laboratories and field trials will be influenced by the economic circumstances of the industry, and will slow down if the underlying profitability declines. On balance, the institutionalization of technical change in the industry through its increasing reliance on off-farm purchased inputs will ensure that productivity growth will continue for some time to come. Some studies indicate that productivity growth will tend to accelerate after the mid-1990s as a result of the coming on-stream of biotechnology innovations. Policy measures to counteract this trend include, for example, measures to slow down the rate of diffusion of cost-reducing or yield-increasing innovations, or limitations on farm size and structural change in agriculture. Such measures increase the resource cost of agricultural programmes to society, although they can be defended on health, social or ecological grounds.

Reducing total factor input may be pursued either by policies which act on the total level of agricultural output (price or quota policies) or by policies aimed specifically at reducing particular categories of inputs. While both sets of measures can be equally effective in limiting agricultural supply, there will be considerable differences in their impact on other agricultural sector goals such as economic efficiency, farm income maintenance and the protection of rural environments. In all cases, policies aimed at restoring balance in any one commodity sector entail the risk of aggravating imbalances in other sectors (e.g. reducing livestock production will reduce demand for feedgrains and aggravate imbalances in that sector) unless such policies are harmonized in the context of a multicommodity approach to policy reform for agricultural adjustment.

It is widely recognised that lower prices for farmers in most countries are an essential element in restoring balance between production and remunerative markets, and that there is no long-term future for an agriculture completely divorced from market trends. The recent policy package of the EC (agricultural stabilizers) is a case in point. Lower prices reduce production or slow down the rate of growth, by encouraging resources to move out of agriculture. A restrictive price policy should also help to reduce the budget and other costs of agricultural programmes and, in the longer term, to reduce agricultural supply. As a method of reducing market imbalances, lower prices have the drawback that farmers' responses are slow and uncertain, and thus the size of the price reduction would probably have to be very severe to have a significant output effect. Farm incomes and the value of assets owned by farmers will suffer and the net result may be the abandonment of agriculture in marginal farming regions unless compensating measures are applied. Market prices have been falling in real terms without appreciable effect on the increase in production because unit costs have also been falling due to technological improvements, structural change and declines in agricultural asset values. The extent of the price cut necessary to compensate for these supply-increasing influences in the short run may not be politically feasible in many countries.

It seems necessary, therefore, that a policy of price reductions should be accompanied by other measures to control production, e.g. limiting the production quantities eligible for support and more generally by marketing quotas or restrictions on input use. *Marketing quotas* have a number of drawbacks, but their main danger is that quota values are incorporated into production costs, create vested interests and the political difficulties of removing them become even more intense. In this case, quotas obstruct rather than ease the adjustment process. It is possible to design quota schemes so that they minimize this effect. For example, producer price levels should not be increased to compensate for the introduction of the quota. If quotas are administratively re-allocated on an annual basis, as is the case under the EC's sugar regime, then no property right in the quota is allowed to develop. Likewise, if quotas are freely transferable, the efficiency costs will be reduced because they can be bought up by the most efficient producers. This would, however, miss the often important objective of keeping small less efficient farms in business, particularly those in disadvantaged regions.

Marketing quotas are at best relevant to a subset of agricultural commodities. For others, *direct restrictions on input use* may be possible. Limiting input use directly has a lesser effect on farm incomes for any given supply reduction but greater economic costs for the wider society. If this option is chosen, the question arises which categories of inputs should be reduced. On the basis of general resource allocation criteria it is most advantageous to reduce the use of inputs which have profitable alternative uses in other sectors. These are mainly the intermediate inputs purchased from the non-farm sector, capital and those members of the farm labour force who have alternative employment opportunities.

Land retirement programmes have a long history, particularly in North America. Withdrawal of land from production has been included as a measure in the 'stabilizers' policy package of the EC. Set-aside land shares some characteristics with the accumulation of reserve stocks, as it is a flexible resource which can be called upon to meet unexpected increases in demand at short notice, although US experience indicates that there is considerable slippage. Again, it is important that such programmes be specifically seen as complementing a restrictive price policy and not as a substitute for it. The possibility of linking land retirement schemes with the achievement of land conservation and more general environmental objectives, e.g. through management agreements with the beneficiary farmers, is an added dimension to be considered in the examination of policy options for supply control.

The taxation of fertilizer inputs has been proposed in some high-use countries on both environmental and supply-control grounds. As an environmental measure it is designed to reduce pollution problems in intensively farmed areas, and as one result it would lead to a more balanced distribution of agricultural production between regions. The impact on production is more problematic, and will depend on the level of taxation imposed and the responsiveness of fertilizer use to changes in its price. It is generally accepted that very high tax rates are required to induce significant declines in fertilizer use.

Increasing production and productivity still an important objective in many countries

Many countries in the region would continue to promote agricultural growth, notwithstanding the general need to lower aggregate growth rates. In some countries this represents a 'legitimate' interest of increasing self-sufficiency (e.g. the USSR), improving overall economic growth prospects (some Mediterranean countries) or recapturing lost shares in world markets. In other countries it is the institutional context that biases policy in favour of higher growth rates, e.g. when individual EC member countries can 'externalize' the costs of increased production because of the common financing mechanism of CAP, or when green rates of national currencies against the European Currency Unit (ECU) are changed in ways which reduce, eliminate or reverse the restraints agreed for the common support prices.

The region is thus faced with the difficult task of reconciling the objective of some countries pursuing improved growth rates while in the region as a whole the growth rate must be lowered. Herein lies the potential for increased conflict and the benefits from negotiated coordinated action become more evident. But there is no getting away from the hard fact that given aggregate consumption one country's production gain is another country's loss, except if the cost of adjustment is forced upon third countries, for which there is not much room anyway. Whatever the solution as to which country adjusts by how much, it is important that relative emphasis be placed on measures increasing productivity and lowering, rather than increasing, the level of price support. This approach is compatible with meeting other objectives, besides increased production, e.g. improved farm incomes, lower consumer prices, and containing budgetary costs. Achieving higher production through increased efficiency in resource use figures prominently among the objectives of the CPE countries. Successful efforts in this direction would be a much needed stimulus to overall economic efficiency, given the significant share of agriculture in total investment and employment.

It is possible that increasing production through increased productivity as conventionally understood (more output for a given amount of inputs) may conflict with other objectives, in particular environmental ones which may require lower input use per unit of land, and social ones, e.g. avoiding the detrimental effects on income distribution of farm concentration inherent in higher productivity farming. The conflict can be resolved, however. If these objectives score high in society's scale of preferences they should be fully internalized in analysing the allocative efficiency effects of policy and in distinguishing more productive from less productive agricultural practices.

Issues related to agricultural technology

Productivity growth based on the application of new technology and better management is a key reason why it is so difficult to control production growth in the effort to redress market imbalances. It is, therefore, natural that in at least some countries the prospect of continued and indeed accelerated gains in productivity may be viewed with mixed feelings. Should governments then be well advised to try to limit the spread of technology in their attempt to control production growth? Certain distinctions must be made here: if the productivity-increasing technology has certain 'external diseconomy' effects, e.g. groundwater pollution from excessive use of nitrogenous fertilizer, its spread will be checked when the costs are internalized (the users of such technology and the consumers of output produced with its aid are made to pay for such external diseconomies) and government policies in this direction will be well taken; other considerations, e.g. short-term ones like slowing down the pace of adjustment or more fundamental ones relating to health risks and ethics, may likewise favour restrictive policies. But overall and in the longer term technology-based increases in productivity are the only sound way of transition to a viable agriculture by increasing returns to factors of production employed in the sector. In this context, European cooperation to promote the advancement and diffusion of agricultural research and technology should continue to receive high priority.

The rate of return to investment in research and technology has been high in the past and will continue to be so. Longer-run economic viability of any activity dictates that high pay-off investment opportunities should continue to be grasped. But the calculation of rates of return must account for all direct and indirect costs of development and application of new technology. Such costs must be defined with reference to society's relative ranking of priorities which, in addition to conventional economic and distributional criteria, will also include 'novel' concerns, e.g. consumer attitudes *vis-à-vis* high-tech based foods, ethical aspects of interfering with biological processes and, of course, environmental concerns.

Policy issues that are emerging as important and will continue to be so in the future include:

— the above mentioned reservations of policy-makers to promoting, or even accepting, new technology having predominantly quantity effects on production in a situation already characterized by excess production capacity;

— the continuous increase in the relative weight of private sector research in the total research effort, resulting from the new opportunities of genetic engineering and the possibility of patenting inventions;

— development of environmentally benign technologies, e.g. low-input farming, including new plant varieties requiring fewer agrochemical inputs;

— possible shifts in comparative advantage *vis-à-vis* the developing countries and the implications for their development prospects;

— related to the above, the increasing regulatory responsibilities of national and international public authorities concerning diffusion and transfer of technological innovations, including the need for strengthened institutional arrangements for safety assessment, risk management and public information.

Farm income issues and policies

Farm income objectives have often dominated other objectives of agricultural policy, particularly in the more prosperous countries of the region. While they will continue to be important in policy-making, there is now increased awareness of the need to define better these objectives so that policy interventions can be targeted more efficiently and to use policy instruments which do not exert detrimental effects on other objectives of policy, as for example sustaining producer incomes by means of 'high' producer prices which further aggravate problems of excess supplies.

An important issue with certain policies aimed at raising farm incomes (e.g. price supports or marketing quotas) is that their beneficial effects are gradually incorporated into higher prices of the agriculture-specific assets (land, quota rights). In this way they tend to benefit (in the form of capital gains) more the farmers and others who own these assets and much less those who rent them (since the effects of policies raise values and eventually also rents of assets). In addition, new entrants need more capital to buy these assets at the inflated prices. Increasing the income of the new entrants would require further increases in support prices, thus perpetuating the vicious circle. Reducing producer prices will reduce the prices of these assets, cause heavy capital losses to those who bought them at inflated prices and bankrupt those who did so on borrowed funds, a phenomenon very prominent up to quite recently, particularly in the USA. In conclusion, some policies aimed at income support raise returns to farm-specific assets, including artificial assets like quotas, and, hence, they benefit farmers and non-farmers in proportion to their ownership of such assets. As such they have unintended effects on farm income distribution and may hinder mobility by increasing the cost to new entrants. Such considerations are important when these policies are discussed as a means of raising or maintaining farm incomes.

The adoption of policies to restrain production growth in many countries, whether in the form of lower prices, marketing quotas or input limitation, will have an adverse effect on farm incomes and farm asset values. Therefore much of the current policy debate is concerned with issues of *decoupling*, i.e.

pursuing farm income objectives by policies which, as far as possible, do not increase production. Greater use of direct income payments to farmers, limits on the quantity eligible for price support as an adjunct to allowing market forces to set the demand–supply balance (e.g. the Production Entitlement Guarantees — PEGs), and the encouragement of dual jobholding are the main policies to be considered. Direct payments can be administered in various ways, in recognition of social need, as compensation for losses imposed by changes in government policies, or as payment for the provision of unpriced services such as environmental conservation, landscape management or amenity provision. Direct payments have been unpopular with farmers in the past, although if the alternative is no payments at all, this attitude will change. They provide the flexibility, however, to target payments in recognition of need or services provided, which in the long run must be a more secure basis for public support for these transfers.

An alternative approach to maintaining farm incomes is to encourage more dual job holding or pluriactivity among farmers. This is an already widespread and increasing phenomenon in both market economies and the CPEs, but with significant country differences (e.g. it is much more pronounced in the Federal Republic of Germany than in the Netherlands). In many CPEs part-time or spare-time farming has grown to be an important source of income of non-farming households, particularly through the exploitation of 'house-gardens'. In the large (state and collective) farm enterprises themselves income-earning opportunities have been greatly enhanced by the introduction of non-agricultural activities, particularly in Hungary, as well as by encouragement of their members to undertake spare-time private farming. The growth of part-time farming has been largely spontaneous to date, although it has been facilitated by government regional development programmes which have assisted in the creation or relocation of non-farm jobs in farming regions. The further extension of dual job holding will depend on increased employment opportunities, both on-farm and off-farm, in rural areas. For example, on-farm activities such as the provision of accommodation services or so-called 'niche' productions might be encouraged. Off-farm employment opportunities can be created through incentives for industrial relocation and the improvement of public infrastructure. As some governments have been less than enthusiastic about part-time farming in the past, the development of an official policy statement recognizing its positive role and contribution would be very helpful.

Increasing importance of farm structure and rural development policies

The transition to a less buoyant agricultural growth environment and less generous support policies compared with the past puts added emphasis on structural measures aimed at creating viable farm units and promoting alternative income sources of the rural population. In parallel, ever increasing national prosperity makes less tolerable the existence of sharp regional differences in wealth and welfare. For example, if GNP grows because of the high-tech revolution, pressures for more equitable sharing of benefits will increase if rural areas in the hinterland are bypassed by progress. The key question then becomes to what extent agriculture can be the base of ever increasing incomes of people working in it. Making it so through structural measures and providing supplementary or alternative sources of income will become an increasingly important issue. More frequently, articulated policy approaches are being developed (e.g. the recent proposals of the EC for the reform of the Community's Structural Funds) which seek to tailor agricultural and other policies to the particular characteristics of widely differing rural areas.

The countries of the region have a long history of measures to promote desirable farm structures and regional development. Policy measures range from land consolidation to facilitation of land mobility and leasing, to physical planning to protect land from non-farm use encroachment and to the creation of regional development boards. All these policies will continue to have a place in total policy-making in the years to come. Relative emphasis will continue to shift towards measures which favour change in the age structure of the farm population in favour of younger farmers, a slowing down of the trend towards the demise of the family farm and associated way of life and promotion of non-farm activities and related income sources in the rural areas. In this latter objective, experiences and trends are similar in both East and West. Indeed, the CPEs have to show some of the most successful experiences in making non-farming activities a major source of income and employment of collective farms, e.g. in Hungary and Czechoslovakia. At the same time, it is thought that the typical size of the socialized farms in the CPEs is too large for efficient management and that this may present difficulties to policy reform aiming at increasing efficiency through self-management.

In the context of current policy reforms in Eastern Europe and the USSR, the issue of farm structures is likely to assume centre stage in the debate of policy options. Obviously, preferences for social control of land resources, in countries in which this is considered important for various reasons, will influence choices concerning farm structures. The pursuit of increased efficiency through the promotion of independent management and market-oriented behaviour of farm enterprises, means that novel forms of farm structure will have to be explored. These would include an increased role for private land ownership and other arrangements aimed at ensuring a long-term and certain framework within which farm enterprises will operate. In this context, note must be taken of trends in many countries which point to an increase in the average farm size and farm concentration. The future role of the large production unit which emerged under socialized farming should be evaluated in the context of such trends, taking account of successful cooperative experiences oriented towards the provision of services to production agriculture.

Need for more integration of agricultural and environmental objectives in policy formulation

In the preceding discussion frequent reference was made to the need for environmental concerns to be considered in agricultural policy-making (and, of course, in the policies of other sectors) not just as an afterthought but as an integral part of it. Increasingly, agricultural, forestry and environmental objectives must be pursued simultaneously in the context of integrated policy approaches. This is particularly important when relative emphasis is placed on regional development which involves the encouragement of agriculture or other activities (tourism, industry) in as yet undeveloped areas.

The more specific issues of the interrelationships between agriculture and environment range from the more obvious ones of pollution and soil erosion to the influences of farming on genetic diversity, and on the amenity value of landscapes to the all-important issues of the effects of environmental protection policies on the objectives of agricultural policy. Depending on market organization and foreign trade regimes, the effects of such policies will be different for producers, consumers and the government. For example, restrictions on some agricultural practices (e.g. use of pesticides) which lead to lower output (a leftward shift in the supply curve) may increase rather than lower aggregate farm income, if prices rise in the presence of low price elasticities of demand, provided such price rises are not counteracted by a surge in lower-priced imports. This is not to say that the incomes of specific groups of farmers will not be adversely affected. The public budget may also benefit if lower output leads to reduced buying into intervention. It will more often be the consumers that will suffer a net welfare loss, as conventionally measured by consumer surplus, a loss which would normally exceed by a good margin any increases in total farm income.

What must be emphasized is the significant diversity of countries as regards conditions in agriculture, including market organization, and the environment, as well as in policy attitudes which reflect different levels of development and economic weights of agriculture. Environmental and economic costs and benefits of natural resource use may not be viewed in the same way by the different countries. Efforts towards integration of policies in agriculture and the environment could benefit from multilateral approaches for two main reasons: first, mutual influences between agriculture and environment can be exerted across national frontiers and benefits of integration may likewise accrue to different countries; and second, action to 'internalize' costs of adverse environmental effects of agricultural activities will raise production costs and, if taken unilaterally, would affect unfavourably the competitive position of the country taking them. Fear of such adverse effects could well lead to action not being taken when needed.

Food policy issues

Food self-sufficiency will most likely continue to figure prominently among the objectives of agricultural policy. It is pursued for a variety of reasons — rural development, the perceived risk of relying on the world market, the desire to conserve foreign exchange, and the protection of food supplies in the face of outside threats to security. However, it is often a costly approach and may not necessarily provide insurance against disruption of food supplies in times of war and similar situations, if there is a high degree of dependence on imported inputs (fuel, raw materials for fertilizers).

Consumer food prices are important instruments of policy in the CPE countries and much of the policy reform debate will continue to be concerned with them. The issue is how to transit towards a system of food prices which play a market-clearing role and reflect more closely social costs and benefits. This involves cushioning the weaker population groups from the immediate welfare losses that removal or reduction of food subsidies involves, while waiting for the compensating welfare gains from improved overall allocative efficiency which will take longer to materialize. Often, however, the implied consumer price increases are very large and, if introduced abruptly, they would spark off inflation in which all population groups, not just the economically weaker ones, would suffer significant income losses which may have to be compensated for. There are, however, no ready answers to the question of how to introduce necessary adjustments without some groups suffering losses. The whole issue of reform of food prices, and indeed of agricultural policies in general, cannot be divorced from the more general adjustments in the overall economic system that CPE countries are attempting. This is particularly so because food subsidies are among the main reasons for large budget deficits which have to be brought under control in the reform process. The recent literature on structural adjustment and stabilization policies for improving internal and external balance may provide some useful insights, since it often analyzes the impacts of large changes in relative prices, devaluation and cuts in government spending. Findings concerning the impact of such policies on different social groups and required compensating measures (e.g. the strengthening of social security systems) could be of particular relevance, in those countries which assign a central role to the objective of preserving a fairly egalitarian distribution of income and preventing the emergence of pronounced inequalities.

Unsatisfied demand for variety and better quality food with a higher degree of processing, better packaging and for more efficient distribution is significant in the CPEs. Much of the effort to improve the agro-food complex will be directed to meeting this demand. In all countries questions of food quality and safety will continue to loom large among the food policy concerns. The issues include how to influence the structure of food consumption away from foods considered harmful to health (animal fats, salt, sugar), to ensure food hygiene through controls, to provide the public with information about the relative importance of the different food hazards (e.g. risks of microbial contamination v. those of food additives), and what should be the policies in the face of demand shifts towards foods coming from 'consumer approved' farm production practices, e.g. without use of pesticides or hormones. Policy-makers will have to decide whether to go for very strict standards (zero tolerance of 'disapproved' technology) or to settle for an acceptable degree of risk, if it is established that certain farm practices entail a real health risk. Choices in this area can have important effects on the rate of productivity improvement in agriculture. Multilateral approaches to the adoption of standards would help avert the inherent change in competitive positions and trade conflicts.

Issues of trade and relations with the developing countries

Much of the impetus for policy reforms in recent years, particularly in the market economies, has come from the 'undesirable' trade-distorting effects which are the result of the pursuit of predominantly domestic goals in agricultural policy. The key question for the future is to what extent countries will accept that agricultural policy must be subjected to the discipline imposed by the need for the orderly development of international economic relations. In practical terms this translates into a quest for policies to meet domestic farm income and related goals by means which minimize distortions of markets.

Avoidance of trade-distorting effects in agricultural policies, implying a greater role for market forces in the determination of production and consumption patterns and trade flows, is by now widely accepted as a valid guideline as to the ultimate goal. However, perceptions differ both as to the acceptable extent of such reform and to approaches regarding implementation. The more radical approach to liberalization advocates the transition to free trade in agriculture. Its proponents would seek to achieve this by means of a gradual and balanced phase-out of all subsidies and trade barriers, with the exception of income-support measures which are not linked to levels of production. Other participants in the global agricultural economy regard it as unrealistic to expect demolition in the years to come of all domestic intervention programmes which restrict trade. Their concerns include the aim of maintaining an acceptable degree of domestic market stability as well as the use of price policy instruments as one of the means for pursuing social and other concerns, such as food security, environment protection or overall employment. Thus their concerns would imply a relative emphasis on the design of improved international trade rules for agriculture which fit domestic realities rather than the other way round. The important thing is that all countries now realize that the deterioration in the conditions of agricultural trade must be halted and reversed, that their domestic agricultural policies can — and often do — contribute to the disarray in international agricultural trade, that there is a real risk of agricultural trade conflicts spilling over into other sectors, and that a coordinated and balanced approach to policy reform would have very real benefits for all participating countries.

The *developing countries* as a group stand to gain from improved market access, mostly by increasing their market share at the expense of industrial-country production of some highly protected commodities, e.g. sugar, beef. The existence of preferential access to the protected markets of some industrial countries by some developing countries will very likely mean that some of the latter countries, as well as those which import these products at the low subsidized prices, may be adversely affected by liberalization even in these commodities. Developing countries would gain, though much less, from improved access for their traditional non-competing export commodities, e.g. tropical beverages, including from more local processing before export, though such gains would be much larger if liberalization were to encompass also the CPEs where the scope for increased consumption is still very considerable.

Developing country exporters of commodities competing with subsidized exports from the industrial countries, e.g. cereals, would gain from the removal of subsidies. The balance sheet of aggregate gains and losses of the developing countries in the cereals sector is, however, somewhat ambiguous. This is so because there is uncertainty as to what would happen to world prices following reduction of protection in the main industrial countries; additionally, the great majority of the developing countries are net importers of cereals. If prices rose, many countries would lose and few (the net cereal exporters) would gain, though all would benefit from the greater stability of world prices following liberalization. The counter-argument is that higher world prices would provide a stimulus and an incentive to the domestic production in the cereal deficit developing countries. These propositions are examined below.

Concerning world prices, it can be expected that reduced protection in the *cereals sector alone* would lower production but not demand and, hence, would tend to increase world prices. If, however, protection were also reduced in the livestock sector, livestock output would decline and along with it the feed demand for cereals. Some studies which simulated the simultaneous reduction in protection in both sectors (e.g. the OECD Trade Mandate Study, OECD 1987) conclude that the impact on world prices of cereals would be non-significant. However, the most commonly accepted conclusion of the trade liberalization studies is that cereal as well as other food prices will rise.

Concerning the potential incentive effect of higher world prices on production in the developing countries, the key question is whether the importing countries could not obtain this effect by taxing subsidized food imports, so that they would save foreign exchange and increase government revenue in the process. In practice, through their own policies, they would be appropriating for themselves part of the farm support expenditure of the industrial countries. The counter-argument is that it is politically very difficult to raise food prices in poor countries by taxing food imports. There is, however, no certainty that increased food prices would be much more acceptable to the consumers just because world prices rose.

There is less controversy as to the direction and magnitude of world price changes of livestock products following eventual liberalization: it is positive and substantial. Developing country importers, particularly of dairy products, would suffer a loss, though actual and, more importantly, potential exporters would gain.

Successful policy reform in the industrial countries would greatly reduce or eliminate their structural surpluses. This would affect adversely concessional exports to the developing countries, both food aid and variously subsidized commercial sales. Given the extreme scarcity of foreign exchange faced by many developing countries, they may not be able to maintain their levels of food imports. In particular, the food security in the low-income food-deficit countries which depend substantially on concessional imports could suffer, especially in those countries which are not significant exporters of other agricultural or non-agricultural commodities whose world prices may rise as a result of liberalization and/or the market access may improve.

In conclusion, the net effect on the welfare of the developing countries of trade liberalization in the particular but all-important case of cereals is much less clear than in the case of other commodities. It must be underlined, however, that the gains and losses resulting from essentially comparative static and partial equilibrium analyses are inadequate indicators of the more dynamic benefits associated with an all-out effort to reduce protectionism. For example, if no progress is made in the cereals sector, the entire effort may well be doomed to failure, meaning no liberalization also in sugar, etc. in which the benefits would be unambiguous and substantial for many developing countries, though not for all of them. There is, however, little doubt that the food security of some low-income food-deficit countries would be adversely affected by trade liberalization which could reduce or eliminate concessional food exports. Safeguarding their food security, at least in the transitional period, would require appropriate policy responses on the part of the international community, e.g. compensatory financing, increased aid and improved market access for the exports, including non-agricultural ones, of the developing countries.

More generally, benefits of the developing countries from trade liberalization would also depend on complementary changes in their own policies, which often discriminate against agriculture, both directly (e.g. low producer prices) and indirectly (e.g. overvalued exchange rates and high protection of other sectors). Some studies indicate that if the developing countries participated in trade liberalization also, their production would increase and this would ease, offset or even reverse, the upward pressure on world cereal prices predicted to result from liberalization in the industrial countries only. These studies are subject in varying degrees to the usual caveats concerning the results of modelling exercises. It can also be expected that if the developing countries liberalized also their policies for their non-competing tropical products (cocoa, coffee, tea) they would collectively suffer significant losses of export revenue (though individual countries may gain) because increased production would depress world prices without significantly stimulating demand in their main export markets in the industrial countries.

Part I

BACKGROUND, HISTORICAL TRENDS AND PROJECTIONS

2 Agriculture in the overall economy

This chapter discusses the role of agriculture in the overall economy, as background to a better understanding of the policy issues and options. The relevant dimensions examined are the relative size of the sector in total gross domestic product and employment, its role in generating incomes for part of the population, its linkages to the other industries of the wider agro-food sector and the total economy and the importance of macro-economic policies for agriculture.

Relative size of agriculture

Agriculture has a relatively small share in the overall economy (as measured by its share in the gross domestic product — GDP) and in total employment in the majority of the countries covered by this study. The study countries, however, span a good part of the world development scene and in some of them agriculture continues to have a considerable weight in the overall economy. The relevant data are shown in Table 2.1. They come from a variety of sources and often they are not homogeneous as to coverage and definitions. In particular, the indicators of shares in employment and of comparative per caput incomes in agriculture and the total economy fail to account for the widely differing rates at which persons working on agricultural holdings allocate labour to, and derive incomes from, non-farm activities. Given this situation in many countries, there is no simple, Table 2.1 straightforward relationship between developments in agriculture (output, prices) and the total incomes of many agricultural households. It is, therefore, very difficult to achieve income objectives in the sector by means of policies which influence primarily agricultural output and prices. The situation varies widely among countries. For example, full-time farming is typical of the farm structure in the Netherlands, while part-time farming is prevalent in the Federal Republic of Germany (FRG). In 1985, 38 percent of all farmers (who were also farm heads) in the FRG were classified as 'with other main gainful employment'. This proportion was between 9 percent and 15 percent in Denmark, France, the UK and the Netherlands.

Large differences also exist as to the extent to which the 'persons working on agricultural holdings' (a concept not equivalent to the employment data of Table 2.1) are fully employed in agricultural work. The agricultural employment shares of Table 2.1 come from the general employment statistics which assign persons to the sector in which they *mainly* work. But many more persons work on agricultural holdings. In the EC-10 there were

Table 2.1 Agriculture in the overall economy: comparative indicators

	Year	Agriculture's [a] % shares in GDP/NMP	Agriculture's [a] % shares in Employm. shares %	Ratio of shares %	Agric. imports % of total imports [b]	Agric. exports % of total exports [b]	Net agric. trade balance [b] in $ million	Current account balance 1985 in $ million
	1	2	3	4	5	6	7	8
1 Belg.-Lux. [e]	1985	2.4	2.9	83	13	11	- 1 089	622
2 Denmark	1985	5.5	6.9	80	11	27	2 958	- 2 708
3 France	1985	3.9	7.5	52	12	17	3 970	749
4 Germany, FR	1985	1.7	5.4	31	13	5	-11 524	13 500
5 Ireland	1984	10.6	15.8	67	13	26	1 450	- 919
6 Netherlands	1985	4.3	5.9	73	15	22	5 462	5 178
7 UK	1985	1.4	2.4	58	13	7	- 6 968	5 155
8 Greece	1984	18.5	27.5	67	15	32	21	- 3 276
9 Italy	1985	4.9	11.0	45	16	7	- 8 814	- 4 132
10 Portugal	1981	8.7	26.6	33	17	8	- 962	379
11 Spain	1985	6.0	17.3	35	12	15	- 85	2 765
12 Finland	1985	7.2	11.5	63	7	5	- 186	- 658
13 Iceland	1983	8.2	10.7	77	11	3	- 82	:::
14 Norway	1985	3.1	7.2	43	6	2	- 762	2 926
15 Sweden	1985	2.9	4.8	60	7	3	- 1 151	- 1 204
16 Austria	1985	3.3	8.2	40	7	4	- 832	- 229
17 Switzerland	1985	:::	6.6	:::	9	4	- 1 843	6 207
18 Albania	1980	34.0	55.9	61	:::	:::	10	:::
19 Cyprus	1984	9.1	26.0 [d]	37 [d]	15	41	19	:::

20 Israel	1984	4.0	6.2 d	100 d	10	14	27	1 099
21 Malta	1984	4.2	5.2 d	65 d	15	6	- 94	...
22 Turkey	1985	17.6	57.1	31	7	31	1 561	- 1 030
23 Yugoslavia	1983	12.1	32.3 d	36 d	10	10	- 140	275
24 Canada	1984	3.2	5.3	60	6	8	2 500	- 432
25 USA	1985	2.1	3.1	68	6	16	11 018	-117 750
26 Bulgaria	1981-5	15.0	26.0	58	7	12	593	- 320 c
27 Czechosl.	1981-5	8.0	18.0	44	9	3	- 1 182	- 80 c
28 German DR	1981-5	8.0	13.0	61	9	2	- 1 558	1 820 c
29 Hungary	1981-4	20.0	26.0	77	9	22	1 115	- 1 287 f
30 Poland	1981-5	18.0	35.0	51	13	8	- 441	- 1 109 f
31 Romania	1981-5	16.0	34.0	47	8	7	173	1 489 f
32 USSR	1981-5	12.0	26.0	46	21	3	- 15 280	3 870 c

Sources: Columns 2–3, countries 1-11, Eurostat, *National Accounts ESA, Detailed Tables by Branch 1987;* countries 12–17, 22, 24–5, OECD, *Historical Statistics, 1960–85,* 1987; Countries 26–32, 1986 (for GDP shares); Countries 19–21, 23, UN, *National Accounts Statistics, Main Aggregates and Detailed Tables 1984,* 1986 (for GDP shares) and ILO, *Economically Active Population Estimates: 1950-80 and Projections: 1985–2025,* 1987 (for GDP shares), and ILO *op.cit.* (for employment). *Columns 5-7:* FAO. Column 8: World Bank, *World Development Report 1987.*

a The data from which these shares are derived generally include forestry and/or fisheries in agriculture. For countries 26-32 the shares refer to material sphere total net material product (NMP) or employment. As such they overstate the weight of agriculture and are not comparable to the shares shown for the other countries. For example, the share in total (material and non-material sphere) employment was in 1985 lower by some 3 (GDR) to 6 (USSR, Poland) percentage points. For the USSR the share of agriculture in NMP is given as 20 percent in 1985 if computed at current prices (ECE 1988: 225).

b Agricultural imports/exports and net balance of crop and livestock products only. Some countries have much higher export shares if forestry and fisheries products are included: Portugal 19%, Finland 37%, Iceland 77%, Norway 11%, Austria 12%, Canada 24%, USSR 6%. All numbers in columns 5-7 are 3-year averages 1984/6.

c Trade balance only (source ECE 1988).

d 1980.

e Columns 2, 3, 4, 8 refer to Belgium only.

f 1986 (World Bank, *World Development Report 1988*)

nearly 13 million such persons, of whom only about 7.4 million had farming as their main gainful employment (data for 1985 and 1986, from CEC 1989a). In Denmark and the Netherlands the great majority of persons working on agricultural holdings were on a full-time basis. By contrast, in Ireland and Italy the persons working on agricultural holdings were almost 2.5 times as many as those shown as having agriculture as their main sector of employment. As a result, every 100 persons working in agricultural holdings provided labour input to agriculture ranging from the equivalent of 41 full-time workers in Italy to between 76 and 79 full-time workers in the UK, Denmark and the Netherlands.

These differences in the weight of part-time farming must be taken into account in policy-making which often has farm income maintenance and improvement as its main objective. For example, on the basis of the general employment statistics (persons working *mainly* in agriculture, forestry and fisheries), it would appear that per caput income in agriculture (net value added at factor cost, average for 1982–6) was 120 percent of the European Community (EC) average in Ireland and only 68 percent in the FRG. However, the situation is reversed (62 percent of the EC average in Ireland, 84 percent in the FRG) when per caput incomes are computed on the basis of the equivalent full-time work years supplied by persons working on agricultural holdings (for data and further discussion, see Table 11.2, Ch. 11).

The relatively small weight of agriculture in the total economy in the majority of the study countries is often at variance with the importance assigned to agricultural policy in the total policy debate. The relationship seems to be inverse: the less important agriculture is as an economic sector, the more is its relative weight in the policy debate. Explanations offered for this phenomenon (e.g. Anderson and Hayami 1986; Hayami 1988) emphasize the increasing willingness of the body politic (reflecting essentially the growing capacity of the non-farm sector to pay the required economic costs and to the diminishing incidence of food prices on household incomes) to ease the burden of adjustment imposed upon agriculture. Such adjustment is made necessary because of the low income elasticity of demand for food in the countries with predominantly domestic-oriented agriculture and the consequent relative contraction of the sector in the process of sustained industrialization and economic growth.[1] Given continuous growth in agricultural productivity, the growth of production capacity tends to surpass that of demand. The internal terms of trade turn unfavourable to agriculture and depress returns to the sector's assets and agricultural labour. Left to market forces, there emerge persistent income gaps between agriculture and the rest of the economy, essentially because of quasi-structural impediments to the flow of labour from agriculture to other sectors, often involving changes in generations. Such sectoral and regional

[1] However, the contraction in the sector's share in total employment and GDP is only partly attributable to the low income elasticity of demand for food. In practice, part of the employment and value added embodied in a unit of farm output has been shifting to the upstream industries (machinery, fertilizers, pesticides). Naturally, this does not happen if such inputs are imported.

income gaps and the rapid flow of labour out of agriculture and the rural areas (rapid, but still too slow for reducing or eliminating income gaps) are often considered socially disruptive and politically undesirable. This sets the stage for support policies for agriculture to ease the pace of adjustment. Such policies, however, tend to become more durable than required to achieve their objective, mainly because they generate rents to particular population groups who, therefore, resist change. The trend for such policies is stronger in countries, particularly densely populated ones, in which industrial development is accompanied by loss of comparative advantage of agriculture. Such countries tend to become net food importers, and this strengthens food security arguments for protection through import restrictions which conceal the economic costs of such policies by not requiring budget outlays.

These characteristics of the European policy scene must be given full recognition in any attempt to understand the past and to explore the future. An examination of policy issues and options cannot be based exclusively on the conventional economic criteria and tools of policy analysis, but it must take into account non-economic considerations of a social and political nature. Indeed, such an examination would be incomplete if it did not recognize the profound influence exerted on policy-making for agriculture (more than for other sectors) by differing perceptions, including doctrinal ones, as to the role that purely economic criteria must be allowed to play in shaping rural societies and farming.[1] Understanding the disproportionate weight of the agricultural policy debate in the advanced industrial countries would require resort to the theory and practice of public choice (see, for example, Hartmann *et al.* 1989) as well as to economics. The latter's primary role is not so much to question the broad objectives but to contribute to their appropriate definition and establish if policies are efficient, i.e. if they achieve the objectives and if they do so at the least cost. With the increasing economic diversification of the rural societies, economics will have an increasing role to play in questioning the pursuit of rural income objectives and regional development by means of predominantly agricultural policies.

Linkages of agriculture

Agriculture is a key component of the much larger agro-food sector, comprising downstream industries (e.g. processing, retailing, catering) as well as upstream ones (e.g. the farm machinery, fertilizers and pesticides industries). These other industries depend in varying degrees for their growth, and sometimes for their very existence, on agriculture. Examination of agricultural policies must, therefore, be enlarged in scope to encompass their effects on these linked sectors and, given inter-industry

[1]   A recent contribution to the policy debate focusing on the agricultural trade confrontation in the North Atlantic area considers that 'deep doctrinal differences, regarding in particular the role to be given to free-market forces in the international trade of agricultural products, have played an important role in the confrontation' (Petit 1988: 195).

linkages, on the wider economy. In such examination, the general proposition holds that the whole economy benefits when resources are allocated to their most efficient uses. If support policies retain in agriculture more resources than would otherwise be the case, economic costs are incurred. There are, of course, good reasons why policy-makers are prepared to accept such economic costs, e.g. food security, environment, preservation of rural communities, etc. However, is it still reasonable to speak of economic benefits to other sectors from agricultural support policies? This question is discussed below, after examining what is the relative size of the agro-food sector in the economy and of agriculture within it.

Data for the USA show that the 'food and fiber' system, including agriculture and all upstream and downstream sectors, accounts for 17.5 percent of the total economy, compared with only 1.8 percent for the farm sector alone (data for 1985, source USDA 1987a). In the seven major countries of the European Community agriculture's share in the aggregate gross value added (GVA) of the economy is 3.1 percent compared with a share of 3.7 percent for the downstream processing industries of food, beverages and tobacco (Table 2.2). These numbers overstate the comparative weight of agriculture in the entire agro-food system of the Community because agriculture includes forestry while the share of the linked industries does not include the downstream sectors of forestry, fibres, distribution and catering nor the upstream sectors of the inputs industry. Further, the valuation of the output of agriculture and the food sector at the high domestic support prices tends to overstate the relative economic weight of the sector. On the employment side, the 'food system' (agriculture, food processing, distribution and catering) of the EC-12 is estimated to account for 20 percent of total employment, or 24 million persons, of which 10 million are farmers, 4 million in food processing and the rest in distribution and catering (Traill 1988). The data available for the European CPEs are not uniform, but they give an idea of the relative weight of the food processing sector (food, beverages and tobacco) in total industrial production and in total employment (Table 2.3).

The data in Table 2.2 indicate that there are wide differences among the EC countries in the relative weight of agriculture in the total agro-food system. At the one extreme, the UK produces 3.2 units of value added in the processing industry for every unit of agricultural GVA. In the Federal Republic of Germany the ratio is 2.1 to 1. At the other extreme the ratios are 0.23 for Greece and 0.57 for Italy. Comparable data for Yugoslavia (Tomic 1989) indicate a similarly low ratio (0.34 in 1985). In principle, such inter-country differences could be explained by a number of interrelated factors, including the degree to which products are processed before they reach the consumer (or are exported) as well as the extent to which the industry processes imported produce (e.g., if for every 100 units produced by agriculture, the industry processes 125 in country i and 110 in country j, then it can be expected that the processing industry will have a higher weight in the total agro-food system in the former than in the latter country, *ceteris paribus*).

Table 2.2 The agro-food system in the European Community

	Shares in Total Gross Value Added at market prices, 1985		Import component of total input of agricultural products [c] in the processing industry, (%), 1980
	Agriculture [a]	Processing industry [b]	
	1	2	3
Belgium	2.4	4.1	...
Denmark	5.5	4.2	11.8
Germany, FR	1.8	3.7	24.8
France	4.2	3.4	12.0
Italy	4.6	2.6	24.1
Netherlands	4.5	3.5	38.2
UK	1.5	4.8	26.2 [d]
All above (EUR-7)	3.1	3.7	...
Greece	17.1	3.9	...
Ireland [e]	10.9	7.5	...
Luxembourg	2.9	3.0	...
Portugal	8.1 [f]	6.3 [f]	25.0
Spain	6.3	5.1	17.6

Sources: Columns 1, 2, Eurostat, *National Accounts ESA, Detailed Tables by Branch, 1988*; column 3, computed from Eurostat, *National Accounts ESA, Input-Output Tables 1980*, 1986.
[a] Agriculture, Forestry and Fisheries; [b] Food, Beverages and Tobacco; [c] Primary and processed agricultural products, of domestic origin and imported, used as inputs in the processing industry; [d] 1979; [e] 1980; [f] 1983.

Table 2.3 European CPEs: shares of the food processing sector, 1985, %

	Share in total industry [a]	Share in total employment [b]
Bulgaria	23.2	10.6
Czechoslovakia	14.2	7.8
German DR	15.5	7.5
Hungary	14.4	12.8
Poland	20.5	9.2
Romania	11.7	6.3
USSR	16.7	...

Source: CMEA data quoted in Jampel and Lhomel (1989).
[a]   It is not known if the shares refer to gross output or to value added.
[b]   Socialized food processing sector only (state and cooperatives).

Concerning the first factor (degree of processing), the generally lower relative weight of the processing industry in the total agro-food system of the Mediterranean countries may reflect the higher share of fresh produce (e.g. fruit and vegetables) in their total consumption and exports compared with the countries of Northern Europe. The second factor (the import component of total produce processed) does not seem to be by itself a consistent explanatory variable of the inter-country variation in the relative share of the processing industry in the total value added by the agro-food system. For Denmark, the FRG, France and the UK, the relationship seems to be in the right direction (the higher the import component the higher the relative share of the processing industry) but the data for Italy, the Netherlands and Portugal do not conform to this pattern. Country specificities, e.g. type of products imported for processing, would probably be additional factors to be considered.

Reverting to the question if farm support policies can be considered to benefit also the other linked sectors, it is to be noted that in a closed economy the case for the existence of some mutual interests is fairly straightforward. It is, however, much less so in a world where both agricultural commodities and inputs can be imported or exported. For example, the interests of the industry processing traded commodities, e.g. cereals or oilseeds, are best served if it has access to the cheapest sources of supply, whether national or imported. Farm support policies which raise prices of their raw materials may harm rather than benefit the industry by increasing its input costs (even though compensatory protection is often provided to the processed product) and restricting consumption of its output. Farm support policies which do not raise prices, e.g. the crushing aid paid to the oilseed processing industry in the EC for its purchases of domestic oilseeds, can be considered neutral for the processing industry. The processing industry may, however, benefit from farm support policies if domestic production provides a more dependable source of raw material supplies. This is thought to be a factor favouring the substitution of sunflowerseed oil for imported groundnut oil in the food processing industry of the EC (Davenport 1988).

The case is simpler for such industries as slaughterhouses, meat and milk processing and more generally the industry processing non-storable and non-traded farm commodities, e.g. sugar crops. In most cases, such industry can only exist if there is local production, while its existence facilitates the implementation of agricultural support policies, e.g. for milk and sugar crops in the EC. In the debate about agricultural policies, it is important to recognize this commonality of interests of the farmers and processors.[1] In between these two extremes lie a large number of agricultural products whose varying degrees of tradeability and perishability are important determinants of the symbiotic or antagonistic relations of agriculture and the downstream-linked sectors. Such relations are also influenced by the structure of the industry (in both the input producing and the processing sectors). Competition policy aimed at controlling monopoly/monopsony power in these sectors, often associated with the spread of conglomerates and multinationals, has, therefore, a role to play in making agricultural support policies more effective.

The case of the interdependence of agriculture and the inputs-producing industry at the national level, is rather more straightforward. This was demonstrated by the difficulties that beset the US inputs industry in the mid-1980s following the agricultural production controls and the difficult farm income situation. In theory, the mutual interdependence need not be so close if the industry is internationally competitive and can export its produce and/or the farmers have access to imported inputs. In practice, however, interdependence is strengthened by the fact that inputs are often sold as a package together with technical and extension services in which the national industry is likely to have a competitive edge over foreign suppliers.

This interdependence of the inputs industry and agriculture is important for the agricultural policy debate since it establishes common interests for farm support, and indeed for support which favours more input-intensive farming, e.g. through shifts in the product mix and technological change. The multinational character of some of the major firms in the industry would, in principle, tend to lessen this commonality of interest at the national level. If, for example, agricultural trade liberalization were to shift some farm production from home to abroad without reducing global output, the loss of input sales at home would be partly compensated by higher sales abroad through the firms' local subsidiaries. Yet, such a relocation of world agricultural production would probably result in a smaller global market for inputs if, as it is likely, it promoted a less input-intensive agriculture at the global level. For example, shifting production of one tonne of grain from Europe to Argentina would have this effect since this latter country uses much less fertilizer than Europe per tonne of grain.

The direct linkages of agriculture to the upstream and downstream sectors

[1]   Dubgaard (1989) referring to Denmark's processing industry (dairy, meat, sugar) states that 'the overall level of activities in the processing industries are determined mainly from the supply side, i.e. by the supply of raw products by the farm sector'.

give, however, only a partial view of the importance of the sector to the rest of the economy and vice versa. These agro-allied sectors are themselves linked to other sectors, so that changes in agriculture are diffused by means of such linkages throughout the economy. An analysis of the complete input–output data for eight EC countries (from EUROSTAT 1986) shows that overall economic growth, which typically raises predominantly final demand for non-agricultural goods and services, will have only a minor impact on agricultural GDP and incomes, as well as on those of the food processing sector, through the inter-industry linkages. The effect could be stronger in terms of per caput GDP in agriculture because vigorous non-agricultural growth could stimulate the outflow of labour from farming. By contrast, an increase in the final demand for the products of agriculture and the food processing industry has a much stronger impact on the rest of the economy through the inter-industry linkages. Given, however, the limited scope for increasing domestic demand, this avenue to stimulating overall economic growth and raising agricultural incomes is not promising (see Ch. 3). This leaves only exports and import substitution as potential stimulants. The scope for the latter is extremely limited, as well as being costly to the economy. The same is true for exports when not based on comparative advantage.

Macro-economic policies and agriculture

Macro-economic policies (through their effects on interest rates, inflation and exchange rates) and general economic conditions have a major impact on agriculture and sometimes surpass agricultural sector policies in importance. At the same time, agricultural policies can be important determinants of macro-economic variables, e.g. through their effects on the balance of payments, particularly in countries with export-oriented agriculture, and on the budget. The general economic situation is an important determinant of the rate of labour outflow from agriculture and, hence, of farm incomes. Likewise, changes in the rate of inflation have been one of the major factors in the strong shifts in the income and asset position of agriculture in the 1970s and early 1980s. When market forces rather than support policies determine prices, increases in the inflation rate tend to raise the relative prices of agricultural products in the short run; the opposite happens when corrective action on the money supply raises interest rates and, for exporting countries, the rate of exchange (Penson and Gardner 1988). In the presence of unchanged market support for agricultural prices, changes in the rate of inflation would affect agriculture mainly via effects on input prices and land values. Thus, in Europe in the 1970s rapid inflation drove up production costs, in turn creating pressures for higher price supports. In the 1980s the sharp fall in inflation was an important factor in reducing input prices, benefiting farmers. In more recent years the decline in energy prices was an important factor in farm income formation, although inter-country differences have been wide.[1] Inflation (in combination with low or even negative real interest rates) contributed to higher land prices and to excessive demands for credit as farmers attempted to acquire more land in order to make capital gains. Inflation also strengthened the competitive position of wealthy farm and non-farm investors in land and led to concentration in land ownership while appreciation of land prices hurt renters and new entrants into agriculture. The reversal of these factors in the early 1980s (lower inflation, higher real interest rates) led to sharp falls in land prices and a substantial deterioration in debt–asset ratios for many farmers. In general, changes originating in the monetary sector represent an added source of instability for agriculture, given the sensitivity of some key variables of the sector to such changes, e.g. asset values and stocks, including livestock numbers. Agricultural policies aimed at reducing instability should pay increasing attention to these factors in addition to the more traditional sources of agricultural variability (see Rausser *et al.* 1986).

Changes in exchange rates, in general caused by factors unrelated to developments in the agricultural sector (e.g. monetary and fiscal policies), lead to trade effects due to changes in competitiveness. Exchange rate changes are particularly important for countries with export-oriented agricultures. They are also important in the EC countries because support prices are denominated in terms of ECUs, and in recent years in 'green' ECUs (the value of the ECU used in the Common Agricultural Policy, which is 13.7 percent above the value of the real ECU). The unwillingness of countries to permit the immediate transmission of exchange rate changes into domestic farm prices has led to an elaborate system of agriculture-specific national (green) rates of exchange which are used to convert the centrally set prices to national currencies. As a result, at the end of 1987 the national support prices, when converted into ECUs at the central exchange rates, varied significantly among countries. They were 7 percent above the EC effective average in the FRG and the Netherlands and 12 percent and 38 percent below it in the UK and Greece, respectively (CEC 1988d). Another result was that by 1988/9 the intervention price of all agricultural products in national currency was on average in the EC-10 12.7 percent above its 1983/4 level, while by then the intervention price in ECU was 1.1 percent below the 1983/4 level, both in nominal terms (CEC 1989b). It is obvious that the existence of the green rates can give a competitive edge to some EC countries over others. To offset their trade effects, an elaborate system of border taxes and subsidies (Monetary Compensatory Amounts — MCAs) has been instituted. The MCAs — because they are only applied to farm output prices but not to farm inputs — have themselves been a source of distorted competitiveness between countries in the EC. The need to

---

[1]   In 1987 the real price of energy in the EC countries declined by over 20 percent in the Netherlands but by only 4 percent in France; that of fertilizer declined by about 20 percent in Ireland and Belgium but increased by 2.3 percent in Greece. Moreover, the differences in the level (as percent of production value) and composition of costs makes for significant differences of the impact of changes in input prices on value added and farm incomes. For example, for each 1 percent fall in input costs net value added would increase by 0.3 percent in Greece and by 1.5 percent in Belgium and the FRG, *ceteris paribus*; net farm incomes would increase by 0.4 percent in Greece and by 3.3 percent in Denmark (CEC 1988c, vol. I).

suppress frontier controls in the process of creating 'Europe 1992' would require policy reform to eliminate the impediments to the free movement of agricultural products represented by the green rates and the MCAs (see CEC 1988d).

Differences in border policies with respect to commodities lead to changes in relative prices following exchange rate changes, e.g. price of some duty-free animal feedstuff imports denominated in US dollars fell sharply since 1985 due to the depreciation of this currency. Exchange rate changes also affect directly the cost of support policies. For example, the increase in the US dollar/ECU exchange rate between 1980 and 1984 (a depreciation of 42 percent of the ECU) effectively reduced the level of protection in the EC and thus budgetary costs, while the reverse was the case from 1985 (depreciation of the US dollar) up to 1988 before the rise in world prices following the North American drought.

3 Demand for agricultural products: historical trends and projections

Demand for food: trends

Food consumption accounts for some two-thirds of total domestic use of agricultural commodities in terms of primary product equivalent in the study countries. Most of the remaining one-third is consumed as animal feed and is, therefore, indirectly influenced by the growth of final demand for animal products. There has been a progressive decline in the growth rate of per caput consumption of food in the study countries as a whole (more accurately, per caput food supplies available for direct human consumption as revealed by food balance-sheet data — see Appendix 2 for explanation of terms) as the levels increased from 3115 calories (kilocalories per person per day) in the early 1960s to 3280 and 3440 calories in 1969/71 and 1984/6, respectively. The growth rate of per caput consumption in terms of farm value declined from 1.5 percent p.a. in the 1960s to 0.8 percent p.a. in the 1970s and to 0.5 percent in the first half of the 1980s. The growth rate of aggregate food consumption declined also due to the slowdown in population growth.

The growth of expenditure on food consumption tends, however, to slow down much less than these figures indicate. This reflects the growing incidence of the processing, transportation and distribution margins in the formation of the retail food prices. For example, estimates available for the USA indicate that in 1985 the farm value of the food consumption basket represented no more than 30 percent of the retail value (10 percent for cereal and bakery products, 45 percent for meats). Ten years earlier these percentages were 40 percent, 19 and 57 percent respectively (USDA 1987b:57). With the increasing popularity of eating away from home (restaurants, work-place cafeterias, fast-food outlets) the spread between the farm value of food and final consumer expenditure is tending to widen further.

There has been significant structural change towards increased consumption of animal products and fruits and vegetables and declines in the consumption of staples like cereals and potatoes. In parallel, the food consumption patterns of the different country groups have tended to converge towards the high animal products pattern, so that differences in the structure of diets in the study countries are today much narrower than in the past (Table 3.2 and Fig. 3.1). These changes reflect above all the general improvement in incomes and living standards, although differences in agro-ecological conditions and food habits still play an important role in the

FIGURE 3.1 - MEAT CONSUMPTION (KG. PER CAPUT)

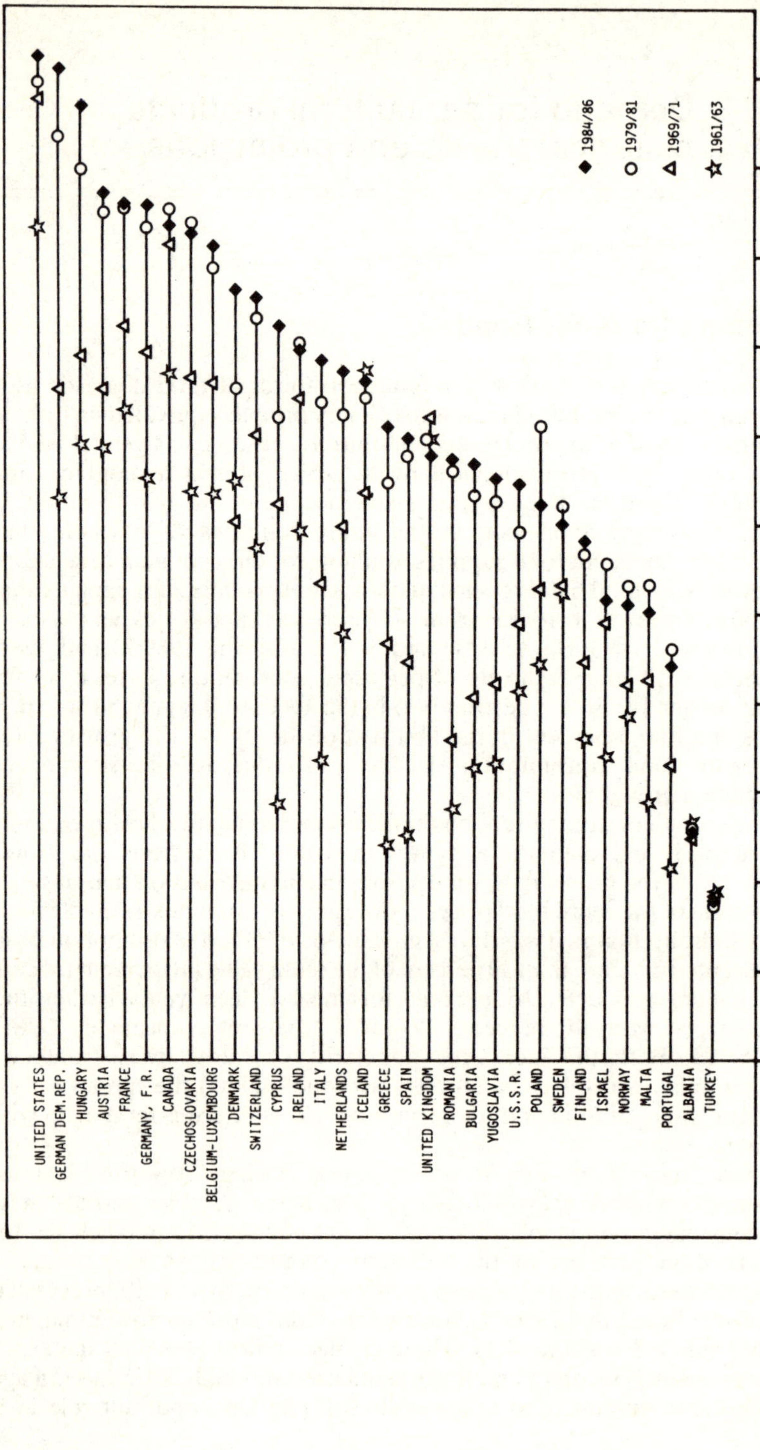

differentiation of diet patterns. Food trade tends to even out those dietary differences that are due to ecological conditions, but this factor is subject to the constraints of foreign exchange availabilities and policy decisions. The influence of these factors is most clearly seen in the consumption levels of typical Mediterranean products in the northern European countries. For example, per caput consumption levels of citrus are 35 kg (fresh equivalent) in the main producing group of EC-South and between 22 and 24 kg in the non-producing country groups of Western Europe (EC-North, Scandinavia, Alpine). These levels contrast with those of under 5 kg in the USSR and Eastern Europe, a level which clearly reflects different food import policies as well as differences in living standards. Similar discrepancies are evident in the per caput consumption of tropical products, e.g. bananas, coffee, cocoa.

Food trade is also an important factor behind the increasing convergence of diet structures towards the livestock-based pattern. This convergence reflects above all increases in the consumption of livestock products in most southern countries and the USSR. Thus, the 2.5-fold increase in the per caput consumption of meat and the 1.8-fold increase in that of milk between 1961/3 and 1984/6 in the EC-South were accompanied by greatly increased net imports into these countries of meat (4.7-fold), milk (7.0-fold) and coarse grains (1.4-fold). Similar developments took place in the USSR, reflecting both the country's priorities in increasing consumption of livestock products, in part based on greatly increased imports of cereals for feed and its improved foreign exchange earnings and low world prices for cereals in the late 1970s and mid-1980s. Wide differences in per caput consumption of meat still persist, though (Fig. 3.1). Health considerations notwithstanding, some countries with comparatively low levels, particularly among the CPEs (e.g. Poland, USSR), consider that it is among their priorities to improve consumption further, both as regards quantity and quality, with relatively more emphasis on the latter.

While the factors discussed above have been the broad determinants of food consumption growth, some other factors have also been present and tended to influence the evolution of this variable, particularly as regards the commodity structure. They include developments in relative prices, health considerations, and shifts in the occupational structure and age composition of the population. These factors have been at play in the well-known case of substitution in consumption of white meats, e.g. poultry meat in North America and both poultry and pigmeat in the EC, for red meats.

Sugar is another example where a general tendency for slowly declining per caput consumption is observed in the higher income countries of the region, mainly reflecting health considerations which encourage substitution by non-caloric sweeteners. This tendency has, however, been greatly reinforced by further substitution due primarily to policy-induced changes in relative prices, particularly in the USA. In this country, high sugar prices due to protection policies in combination with low (near world market) prices for maize and other factors (e.g. the relatively large share of industrial use of sweeteners which favours their use in liquid form, the advent of improved enzyme technology) have led to drastic substitution of corn syrup for sugar (for a description and analysis of the USA sugar policies see Maskus 1989). Per caput consumption of sugar in the USA declined from

50 kg in 1969/71 to 32 kg in 1984/6 (raw equivalent). In parallel, the share of substitutes (mostly high fructose corn syrup — HFCS or isoglucose — and glucose corn syrup) in the total caloric sweetener consumption increased from 24.5 percent in 1975 to 51.2 percent in 1985. Substitution has tended to level off in recent years, mainly because HFCS has captured most of the industrial sweetener market, although the prospect of crystallized HFCS production could rekindle the process. The US experience is an example of predominantly policy-induced changes in food consumption patterns at the expense of imported food and in favour of a domestic substitute. In Europe substitution has been much more limited, mostly because maize prices in relation to those of sugar are much higher, sugar itself is in surplus, industrial consumption is a smaller component of total sugar use and also because EC policy has been to discourage corn syrup production.

The evolution of food consumption also reflects the food price policies pursued by the different countries. Three broad patterns can be distinguished in the region. In most countries of Western Europe domestic prices have been kept well above those that would prevail in the absence of import restrictions. The North American pattern has been characterized by virtual absence of border restrictions for the export oriented commodities and their domestic prices were generally allowed to follow the evolution of international prices. Other commodities, however, have been subject to intervention with notable effects on domestic prices, e.g. sugar and dairy products.

In the CPEs the pattern has been opposite to that of Western Europe. In most CPE countries consumer prices have been kept low and stable until quite recently. The term 'low' here means well below those that would have prevailed in the absence of state subsidies which can be quite high. It is estimated that in the German DR for every 100 marks spent on food in 1985, the state provided around 78 marks in subsidies (Köhler 1989: 177). In the USSR the subsidy element on livestock products is even higher, as the following data show:

USSR, 1985	*Retail price (in rubles/Kg)*	*State expenditure (in rubles/kg)*		*Total*
		Total	*of which subsidy from the special budget account for price regulation*	
Poultry meat	2.57	2.92	(1.13)	5.49
Beef and veal	1.75	5.42	(4.03)	7.17
Mutton	1.42	4.86	(3.31)	6.28
Pigmeat	1.84	3.51	(2.12)	5.35
Milk	0.25	0.45	(0.30)	0.70
Butter	3.38	8.43	(6.30)	11.81

Source: Semenov 1987.

If the numbers in the last column are really the aggregate cost of making available to the consumer 1 kg of these livestock products,[1] it means that the consumer pays directly only a fraction of the total cost and pays for the balance indirectly through taxes, higher prices for other goods and, in some cases, lost time in food lines.[2] Such policy tends to favour food consumption at the expense of other goods and services that receive relatively less subsidy or no subsidy at all. It had some justification when food consumption levels were low and, when accompanied by low producer prices, it helped transfer resources out of agriculture for industrial development. It may be counterproductive at the current high consumption levels of staple commodities by encouraging wasteful use of food and maintenance of poor quality.[3] The related argument that low food prices benefit the poor more than the rich, at least in the short-term, is correct by itself but has probably little relevance if income distribution is fairly egalitarian. In any case, the maintenance of low food prices for rich and poor alike is a very costly and inefficient means of achieving objectives of equity and income redistribution (see FAO 1987). In the USSR, it is thought that the concentration of subsidies on livestock products tends to accentuate social inequalities, since economically better-off families consume more of these 'luxury' products and hence benefit more from the subsidies than lower income families (Aganbeguian 1987:200).

Notwithstanding the strong economic justifications in favour of significant reductions or removal of food subsidies, implementation of such a policy is likely to face strong resistance because food subsidies are such an integral part of the economic system of the CPEs. Köhler (1989: 177), for example, cites the low subsidized prices for basic foods, which remained unchanged for the last 25 years in the GDR, as a significant contribution to the climate of security in society and an important achievement of socialism. At the same time it is recognized that food subsidies as well as those for rents and services to the population are not a small burden on the GDR state budget (Schilar, 1989:31). If food subsidies are reduced or removed consumers will suffer an immediate loss of income while gains from the more efficient use of resources will take time to materialize. Implementation may therefore be very difficult without some form of compensation involving income increases, at least for the population in the low income groups. However, large and abrupt price increases for food, leading to higher inflation rates,

[1]   Data for 1983 indicate the following 'average production costs' in rubles/kg: beef 4.75, pork 3.25, milk 0.42, butter 8.18 (data from *Finansy SSSR*, no. 4, 1985, given in Cook 1987: 15). Average production cost may, however, not be the all-inclusive cost for bringing to the consumer one kg. of these products.

[2]   Of central interest to the policy debate is the extent to which food subsidies succeed in keeping food prices low, e.g. in terms of work hours (percentage of average wage) required to buy a representative basket of food products. In this context, the persistence of relatively high shares of food in total private consumption expenditure in most CPE countries, notwithstanding food subsidies, is telling.

[3]   'Les prix minorés des produits agricoles permettaient de redistribuer les ressources de l'agriculture à l'industrie, et ainsi de garantir l'alimentation de familles indigentes. Dans les conditions actuelles, ces prix minorés freinent notre développement et favorisent le gaspillage' (Aganbeguian 1987:148). It is estimated that out of a total annual bread production of some 40 million tonnes, around 5 million tonnes is used for animal feed (International Wheat Concil 1989).

would have adverse welfare effects on the whole population. In such a case, provision of full compensation through higher money incomes may not be feasible, since it could jeopardize the whole reform effort (for further discussion, see Ch. 10).

Demand for food: projections

The reasons responsible for the progressive decline in the growth of food demand (mainly the prevalence of high average consumption levels, though not in all countries and commodities, and the slowdown in population growth) will continue to be present in the future. As a consequence, a further slowdown in food demand growth may be expected. For the region as a whole the annual growth rate for the period 1984/6–2000 may be only 0.9 percent, compared with 1.3 percent in the preceding 15 year period 1971–86 (Table 3.1).

This projection of aggregate growth of food demand is derived from projections for individual countries and commodities taking into account projected population and incomes. Not all countries will have low growth rates in demand since the region includes some countries which have high growth rates of population and/or comparatively low per caput consumption of some commodities (for country data see the statistical tables in the Appendix). Thus, while population growth is projected to be zero or negative in a number of countries of Western Europe (the FRG, Sweden, the UK, Switzerland, Austria), there are also countries with projected population growth rates close to 2.0 percent p.a. (Turkey, Albania). In between are the two largest countries (USA, USSR) which together account for 50 percent of the total population of the study countries and have projected population growth rates of 0.8 percent p.a.

The slowdown in population growth accounts for one half of the fall in the growth rate of aggregate food demand. The other half is due to the natural slowdown that follows the achievement of high levels of consumption (declining income elasticity of demand). Measured in terms of primary product equivalent, the income elasticity of the demand for food is very low in most countries of the region, though exceptions exist, particularly in the non-EC Mediterranean region, and in some CPEs, particularly for meat. In these countries the growth rates of incomes play a more decisive role for the evolution of food demand than in the more prosperous countries of the region. In the CPE countries the evolution of food consumption may be decisively influenced by policy reforms directed at changing the process of consumer price formation, generally leading to an increase in the relative price of food. In these same countries food import policies will also play a more decisive role than in other countries in the evolution of food consumption, particularly as regards commodity composition (see Ch. 4 for discussion of factors affecting food imports into the USSR).

The above-discussed growth rates of food demand refer to the farm value of food, i.e. direct human consumption of food products expressed in primary product equivalent and valued at international prices, as explained

Table 3.1 Projections of population and food demand

	Population				Food demand				Calories/caput/day	
					Total		Per caput			
	1985	2000	1970–85	1985–2000	1971–86	1984/6–2000	1971–86	1984/6–2000	1984/6	2000
	million		.... Growth rates percent p.a							
EC-north	206	209	0.2	0.1	0.7	0.2	0.5	0.1	3360	3430
EC-south	116	122	0.7	0.4	1.5	0.8	0.9	0.4	3440	3550
EC-12	322	331	0.4	0.2	1.1	0.4	0.6	0.2	3385	3470
Scandinavia	18	18	0.3	0.0	1.2	0.2	0.8	0.2	3100	3160
Alpine countries Non-EC	14	14	0.1	0.0	1.1	0.1	0.9	0.2	3425	3480
Mediterranean countries	81	101	1.8	1.5	2.5	2.1	0.8	0.6	3240	3380
Total Western Europe	435	464	0.6	0.4	1.2	0.7	0.6	0.2	3350	3440
North America	264	298	1.0	0.8	1.1	0.9	0.1	0.1	3620	3680
Total Western Europe + North America	699	762	0.8	0.6	1.2	0.8	0.4	0.2	3450	3540
Eastern Europe	112	120	0.6	0.5	1.6	0.9	1.0	0.4	3460	3530
USSR	278	314	0.9	0.8	1.6	1.3	0.7	0.5	3395	3490
Total CPEs	389	433	0.8	0.7	1.6	1.2	0.8	0.5	3415	3500
All above	1088	1196	0.8	0.6	1.3	0.9	0.5	0.3	3440	3520

in the Appendix. Measured in this way (which is appropriate for examining the effects of the growth of food demand on farm output), per caput food demand would increase by some 5 percent over the period 1984/6–2000 in the region as a whole. In terms of calories, however, it would increase by half as much over the same period. This is so because the process of diversification of food consumption towards those commodities which provide fewer calories per unit of farm value would continue, though at a slower pace than in the past as more and more countries attain high (near saturation) consumption levels also in these commodities. In practice, the farm value price per calorie, which increased by some 5 percent over the past 15 years, may increase by a further 2.5 percent over the projection period. The projected changes in the commodity composition of food consumption are shown in Table 3.2.

As noted earlier there is an increasing gap between the growth rate of food demand in terms of farm value and actual expenditure on food. This reflects the increasing role of processing, packaging and distribution margins in the formation of the retail price. The US data mentioned earlier quantify and demonstrate the extent of these margins. Such trends would continue, in particular the tendency for an increasing share of expenditure on food to be spent on food consumed away from home (Blaylock and Smallwood 1986). Additionally, expenditure on food may be further stimulated by consumer willingness to pay higher prices for food perceived to be of superior quality, e.g. organically grown vegetables, additive-free processed food, free-range eggs and meat, particularly in the higher-income countries of the region. It is also likely that policy reform in the CPEs, including consumer price reform, would tend to increase the value added component of the processing and distribution sectors in the total final cost of food. This assumes that such reforms would be in the direction of meeting latent demand for better quality food as well as for restaurants and related services.

All projections are subject to a margin of error and other projections show different results. For example, one study for the USA (Myers *et al.* 1987) projects domestic demand for food, expressed in primary product equivalent, to grow at about 1.2 percent p.a. through the year 2000, against a growth of 0.9 percent p.a. in this study. In a more recent assessment Gardner (1989) expects domestic demand in the USA in the 1990s to expand at about the population growth rate. Another report (CEC 1988b; see also Traill 1988) expects that EC food demand (in quantity terms) over the next 20 years will be static or even fall, against a 0.4 percent p.a. growth assumed in this study. Particularly uncertain are the food demand projections for CPE countries. In spite of the reported high calorie levels, 45 percent of the calories are supplied by basic food crops against only 35 percent in Western European countries with comparable calorie supply levels. This suggests that there is still considerable room for changes in dietary composition, and many CPEs consider that they should increase meat consumption considerably, notwithstanding the significant gains already achieved in some countries, at least according to the national average data in the FAO food balance sheets (for country data see Appendix Table A.3). These data show generally higher per caput availabilities than in some high income Western countries, for example in the Scandinavian group.

Table 3.2 Food consumption (kg. per caput; in primary product equivalent)

	Basic food crops*			Sugar (raw equiv.)			Veg. oils, oilseeds (oil equiv.)			Fruit and vegetables (fresh equiv.)			Tropical beverages†			Meat (carcass weight, excl. offals)			Milk and products excl. butter (fresh milk equiv.)		
	69/71	84/6	2000	69/71	84/6	2000	69/71	84/6	2000	69/71	84/6	2000	69/71	84/6	2000	69/71	84/6	2000	69/71	84/6	2000
EC-north	221	207	195	43	40	37	11	13	15	232	237	250	7.6	8.8	9.2	73	84	87	220	240	248
EC-south	242	226	213	29	30	31	18	21	22	402	406	442	3.2	4.8	5.7	46	68	76	154	219	230
EC-12	228	214	201	38	36	35	14	16	18	291	297	321	6.1	7.4	7.9	64	78	83	197	232	242
Scandinavia	192	184	178	46	43	41	11	11	12	112	143	170	13.2	13.3	13.6	46	55	58	291	340	336
Alpine countr.	199	171	152	45	41	44	12	14	15	277	294	313	6.2	9.0	10.1	71	90	92	247	290	289
Non-EC Medit.	269	277	251	24	31	38	9	14	19	258	270	302	1.4	2.5	2.7	29	34	38	98	108	115
Total Western Europe	232	223	210	36	36	36	13	15	18	278	286	311	5.7	6.7	7.1	58	69	72	187	216	219
North America	161	177	180	50	34	33	17	23	23	196	238	259	8.4	7.1	6.6	105	109	112	245	239	244
Total W. Eur. + N. Amer.	206	206	198	41	35	35	15	18	20	248	268	291	6.6	6.9	6.9	75	85	88	208	225	229
E. Europe	308	279	259	36	40	41	8	9	10	166	213	249	2.1	3.1	4.0	57	79	83	185	212	231
USSR	336	284	267	42	47	48	7	10	13	134	170	214	1.0	1.8	2.4	48	65	71	192	171	180
Total CPEs	327	283	265	40	45	46	8	10	12	143	183	224	1.4	2.1	2.7	51	69	74	190	183	194
All above	249	233	222	41	39	39	12	15	17	211	237	266	4.8	5.2	5.4	67	79	83	202	210	216

* Cereals, pulses, potatoes and other root crops.
† Cocoa and products, coffee (beans equivalent) and tea.

It is, however, possible that the data for some CPE countries overstate food supplies available for final human consumption. In some countries, inappropriate methods of reporting production statistics result in overestimation of available supplies (Boulatov 1989; Shmelev and Popov 1989; 30-31). Additionally, insufficient allowance is made for food losses. The two are interrelated: production overestimates make losses (and probably some uses of agricultural products such as feed) appear to be higher than they really are. For the USSR, food losses were estimated recently by President Gorbachev to account for some 20 percent of agricultural production, with peaks of 30–40 percent for some products (Ambassade de l'URSS — Berne, 1989)[1]. Lukinov (1989) estimates that losses of agricultural produce 'nowadays amount to one fourth of agricultural output'. Higher estimates of losses for some products are given by Tikhonov (1988). The International Wheat Council (1989) gives loss estimates of some 20 million tonnes for grains and some 2.4 million tonnes for meat, which are around 10 percent of total domestic use of these products. These loss estimates are about the same as those underlying this study for grains but far greater for meat. These estimates can be compared with figures for the EC-10 (some 2 percent for grains, nearly zero for meat) and much lower ones for grains and also nearly zero for meat in the USA. If, as is likely, per caput food availabilities are in reality significantly below those shown in the food balance sheets, particularly for meat, potatoes, fruit and vegetables, then this would explain in part the apparent excess demand for food which is reported for the USSR. A downward adjustment in the data of per caput food availability and, eventually, production for the base year, while retaining the projection for the year 2000, would result in a higher growth rate for these two variables than shown in this study. This fact need not, however, disturb the absolute levels of the projected demand–supply balance. If efforts to reduce food losses are successful, it would be possible to meet part of the demand increases without corresponding increases in farm output or imports. The above mentioned loss estimates of the IWC are equal to some 60 percent of net imports of cereals and far above net imports of meat.

Attempts at explaining the excess demand for food in some CPEs, particularly meat, usually emphasize also the excessive growth of money incomes not fully matched by corresponding increases in supplies (see, for example, Aganbegyan 1989), the maintenance by means of subsidies of low food prices, already noted earlier, and shortages in non-food consumer goods on which expenditure could be diverted. To the extent that these problems are tackled in the future and with the relative shift of demand from quantity to quality, they would tend to dampen the growth rate of food demand in terms of primary product equivalent. Cook (1987: 16–17) considers that 'even assuming relatively inelastic demand for meat in the

[1] Treml (1986) considers that the FAO data overstate per caput food availabilities (calories) in the USSR, although such data include an allowance for calories provided by alcoholic beverages (some 5 percent of the total) which Treml does not consider as being food. An interesting analysis and critique of the USSR's estimates of meat consumption is given by Chernyak (1988).

USSR, an abandonment of food price subsidies in the USSR would probably go a long way toward eliminating the Soviet market shortages of meat'. While the statement is correct, the argument has to be treated with caution. The gist of the issue is how to bring about price policy reforms and establish market balance, while minimizing adverse effects on the food consumption and nutrition of the poor.

Demand for feed: trends

As noted, most of the domestic use of agricultural products other than for final food consumption, is represented by animal feed. Comprehensive estimates of feed use are available only for cereals and oil meals. In the case of cereals and for the region as a whole, feed accounts for 63 percent of total domestic use (including on-farm use). Over the last 25 years all country groups and nearly all countries moved to a situation where the feed sector accounts for more than 50 percent of total use of cereals. This was not so in earlier years. This convergence is in practice a mirror image of the convergence of the diet composition towards the livestock-based model noted earlier. It was achieved primarily through the catching-up of the lower-income country groups.

This predominance of the feed sector in total demand for cereals should be seen in the perspective of the more general move towards increasing dependence of the livestock sector on concentrates and away from roughages. This trend reflects essentially the shifts in the composition of livestock output towards products from monogastric animals (pigmeat, poultry meat and eggs) which favoured the increased use of concentrates. These trends were reinforced by the spread of intensive grain-based feeding systems for poultry and pigs as well as some ruminant livestock, such as dairy cows and beef-cattle (FAO/CCP 1988). There are wide differences among countries as to the extent of substitution of concentrates for roughages in total feed supplies. The data on total feed supplies (including concentrates and roughages) are notoriously inadequate. Estimates for the USSR indicate that cereals account for 25 percent of the total feed supplies (cereals and roughages in feed units of oats equivalent — Diamond *et al*. 1983: 162). Some OECD estimates indicate that in the early 1980s the contribution of concentrates to total feed supplies (in terms of metabolizable energy) ranged from 50–60 percent in the USA, Canada, Spain and Portugal to only 20 percent in Ireland. Within the concentrates group the shares of cereals was predominant, but also here there is wide variation: 86 percent in Canada and only 17 percent in the Netherlands and 28 percent in Belgium, although shares of two-thirds were typical of most other European countries. These wide differences reflect above all the fact that, within certain limits of technology and animal nutrition, individual feeds are substitutes for each other and, therefore, the composition of the least-cost feed mix is highly sensitive to their price ratios. Thus, the low cereal component in the concentrate feeds use in the Netherlands and Belgium reflects, among other things, the price advantages enjoyed by imported substitutes near the ports of entry (OECD 1989b: 45).

In the 1960s the growth of feed use of cereals was very strong and underpinned that of the cereals sector. During the decade feed use absorbed 83 percent of the increment in cereals production. This trend lost momentum during the 1970s so that further increases in feed absorbed only 45 percent of the increment in cereals production. The dynamic element that underpinned the cereals production growth of the region in the 1970s was the foreign trade sector; net exports to the rest of the world increased three-fold during the decade and absorbed another 45 percent of the increment in production of the region as a whole. In the first half of the 1980s there was further slowing down in the use of cereals as feed and this coincided with the virtual stagnation of the net exports of cereals from the region to the rest of the world. These developments, which also reflected and were reinforced by the tighter monetary policies of the early 1980s (Rausser et al. 1986), led to the well-known problems of government stock accumulation in the mid-1980s and have been a key element in the debate, both national and international, concerning required policy reforms.

Feed consumption of oilmeals grew in most country groups faster than that of cereals and moreover the high growth rates were maintained during the 1970s and slumped only in the first half of the current decade. The notable exception was the USSR where oilmeal consumption per unit of livestock output remained very low (for data see UN/ECE 1988:227). It is thought that the low protein content of feed rations has been one of the reasons for the inefficient use of cereals in feed (Aganbeguian 1987:160).[1] The developments in the use of cereals and oilmeals as feed reflect the growth, the species and output composition of the livestock sector, the relative prices of the different feed products, noted above, and their relation to the prices of the livestock products. They also reflect technological progress in the feed–livestock sector and, particularly in the CPEs, balance of payments constraints and import policy.

A basic factor behind the maintenance of the growth of feed demand for oilmeals but not for cereals in the 1970s was the faster growth of the pig and poultry sectors (which require higher protein content in feed rations) compared with the rest of the livestock economy. In Western Europe this factor was greatly reinforced by the penetration of the EC feed markets by cereal substitutes (or non-grain feeds) some of which, e.g. cassava, required oilmeal supplements to increase the protein content of feed rations. EC imports of cereal substitutes increased rapidly from 3.7 million tonnes in 1970 to a peak of 16.2 million tonnes in 1982 before declining to between 13.6 and 15 million tonnes in the subsequent years (Schmidt and Gardiner 1988). These largely imported feeds substituted for both cereals and roughages (mostly pastures, hay and root crops). The EC policies were instrumental in encouraging these developments by keeping cereal prices high in relation to those of substitutes, both imported and domestically produced, e.g. pulses. A recent EC study stresses how these 'disharmonies'

---

[1]   Other factors considered to contribute to the low feed conversion ratios in the USSR include the low share in the cattle herd of animals especially bred for beef production, the fact that much of the grain is fed to the animals in unprocessed form and the losses in feed quality resulting from inadequate harvesting and storage technologies (Diamond *et al.* 1983: 166).

in agricultural policy measures have been responsible for the fact that '... EC livestock production growth has been achieved with nearly no increases in feed grains use' (CEC 1988e: 1/8). The US protection regimes for sugar and ethanol further reinforced these developments by encouraging the supply and export to Europe of such substitutes (mainly corn gluten feed), which are by-products of both ethanol and high fructose corn syrup.

The sheer weight of the feed sector in total cereals use in the study countries is an important factor in cereal market developments, both domestic and international. In particular, feed demand for cereals is considered to be more responsive to price changes (has higher price elasticity) than the demand for food. This reflects the already noted substitutability of the different feedstuffs as well as the higher price elasticity of food demand for animal products than for cereals. In times of supply shocks, e.g. shortages due to weather fluctuations and consequent rises in prices, the contraction of the feed demand for cereals and, eventually, of the livestock sector itself, contributes to the mitigation of price rises. As a consequence, market price instability is less pronounced and direct consumption of cereals is less affected, than they would have been if the bulk of demand for cereals were for food purposes.

These considerations can be a contributing factor in world food security, in particular in lessening price instability for food importing developing countries. For this mechanism to work, however, changes in supply must be allowed to influence the price actually paid for cereals by the livestock sector. This is not always so if domestic markets are shielded from price movements in the world markets, or if domestic supply changes are compensated at nearly unchanged prices through resort to foreign markets. The differential behaviour of more or less shielded markets can be seen from a comparison of changes in feed use of cereals in the USA and the EC-10 following the decline in world cereal production in 1972 and the sharp rises in world prices in subsequent years. In the USA feed use of cereals in the three-year average 1974/76 was 32 million tonnes (or 22 percent) below that of the preceding three years 1971/73. In the EC-10 the decline was only 1.2 million tonnes or 1.7 percent.

Demand for Feed: projections

Demand for feed is derived as a function of livestock production. The livestock projections underlying the present discussion of feed demand are presented in Chapter 4, where it is explained that these projections can be defined only for two groups of countries: the CPEs and all the other countries lumped together with the developed countries not covered in this study (Japan, Oceania, South Africa). Analysis of the feed–livestock complex is hampered by unavailability of complete feed balance-sheet data. In what follows, therefore, only some rough trend analyses and projections are offered for cereals and oilcakes/meals. The amount of feed (cereals, oilcakes) used per unit of livestock output has tended to stabilize in recent years. It is assumed that this near stability of the coefficients will continue to prevail over the next 15 years. By implication, feed demand for both cereals and oilmeals would grow at approximately the same rate of livestock production (Table 3.3).

Table 3.3 Feed coefficients and growth rates

	Kgs of feed per kg livestock output *				Growth rate of feed		
	61/63	69/71	79/81	83/5	61-70	70-85	83/5–2000
Cereals							
Market economies†	2.32	2.63	2.37	2.36	3.6	1.0	1.0
CPEs	1.63	2.65	3.12	2.91	10.1	2.5	1.3
Total	2.12	2.64	2.60	2.53	5.2	1.5	1.1
Oilmeals							
Market economies†	0.26	0.32	0.44	0.44	4.9	4.2	1.0
CPEs	0.12	0.16	0.20	0.20	6.9	3.3	1.3
Total	0.22	0.27	0.37	0.37	5.4	4.1	1.1

*   Livestock output = meat + eggs + milk × 0.2.
†   Western Europe and North America plus the non-study developed countries (Japan, Australia, New Zealand, Republic of South Africa).

These projections may appear conservative but there are indications that technological progress in the future would increase productivity (reduce the amount of feed required per unit of livestock production) by as much as 2 percent p.a. in the poultrymeat sector, 0.6 percent p.a. in that of pigmeat but only 0.2 percent p.a. in the beef and dairy sectors (US Congress, Office of Technology Assessment, 1986). A study for the USSR projects feed use of cereals between 1980 and 2000 at a rate which is somewhat below that of meat production, the ratio between the two growth rates (elasticity) being 0.85 (Diamond *et al.* 1983). The EC Commission are pessimistic concerning growth of feed demand until 1994 which they project to remain constant in total but to decline for cereals (CEC 1988c). A recent study on 'Agriculture Beyond 2000' in the Netherlands foresees a strong trend towards substitution of roughages for purchased feedingstuffs (Douw *et al.* 1987). The oilmeals projections for the CPEs may be on the low side given the widely-held belief that feed productivity could be increased significantly by increasing the protein content of feed rations. A higher growth rate of oilmeal use would, however, imply a lower growth rate of the cereals used for feed. Depending on the technical rates of substitution and relative prices it could be economically rational to substitute part of the imports of feedgrains by imports of oilseeds, notwithstanding the higher price of the latter. This course is being increasingly followed in recent years (Johnson 1989).

The achievement of more efficient use of both cereals and oilseed feeds in the CPEs would require continued emphasis on greatly improving alternative sources of feed (grasslands, roughages), including more emphasis on increasing protein feeds from sources other than oilseeds.[1] It would also involve wider diffusion of modern, more feed-efficient technology already used in part of the livestock production system. The projections study for the USSR mentioned above carried out sensitivity analysis in this respect. This analysis indicates that the ratio of the growth rate of cereals use for feed to that of meat production could be reduced further to 0.67, following wider adoption of such technology. This result assumed that the share of meat production originating in enterprises using this technology — production in 'complexes' — increased from 14 percent in 1979 to 48 percent in 2000. However, the adverse environmental effects of heavy concentrations of livestock (especially pigs) in large industrial units should not be underestimated. This is a factor that would probably militate against the pursuit of higher feed efficiency by these methods. All things considered, therefore, there is considerable scope for increasing feed use efficiency in the USSR. In recent years only modest progress has been made in improving these rates (IWC 1989). Whether faster progress will be made in the future depends on the success of policy reforms currently under way, as discussed in subsequent chapters.

Aggregate domestic demand

Food and feed use of farm commodities account for some 90 percent of all domestic uses of agricultural products, including imported ones. The balance is used for seed and raw materials for the production of non-food final products (fibres, tobacco, vegetable oils, grains for fuel alcohol, etc.). Projecting demand for industrial non-food uses is much more difficult than for food or feed. This is because some of the commodities concerned (fibres) are subject to competition from synthetic products as well as to substitution by imported manufactures, e.g. demand for mill consumption of cotton depends on trade policies concerning imports of textiles. Finally, the use of some food commodities for the production of non-food industrial products depends to a great extent on support policies. In recent years the debate concerning such uses of surplus food commodities has received new impetus from the policy impasses related to the control of excess production. Prospects in this direction are, however, not encouraging (see Box 3.1; also Neville-Rolfe and Caspari 1987).

Demand for non-food industrial uses of all agricultural commodities is projected to grow at 1.1 percent p.a. for the region as a whole, which is roughly the same as the growth rate of the 15 years to 1985. The aggregate domestic demand (all uses, all commodities) projections are shown below.

---

[1]   The USSR has been emphasizing production of single-cell proteins (with planned production of 3 million tonnes) as a substitute for imported soybeans (OECD 1989a: 88).

Table 3.4 Growth rates of aggregate domestic demand (percent p.a.)

	1961–70	1970–85	1983/5–2000
Western Europe and North America	2.0	1.2	0.8
Eastern Europe and the USSR	3.7	1.5	1.2
Total	2.6	1.3	1.0

The growth rate of projected aggregate demand for the CPE countries may appear to be on the low side in the face of reported excess demand and shortages in some countries. These growth rates result from the comparison of the projected and base-year demand levels. If, as discussed above, the base year statistics overstate current availabilities, and hence implied consumption, the growth rate of demand required for attaining the projected absolute levels will be higher than indicated here. Other factors that instead may play a role in restraining demand growth in some CPE countries include higher food prices following price reform, more emphasis on quality rather than quantity, increased availability of non-food consumer goods and services, draining excess money liquidity from the system and improved feed-use efficiency in the livestock sector. It is to be noted that the latest World Bank projections for the USSR (World Bank 1989) have significantly higher projected demand for cereals but projected production nearly equal to that of this study, with the result that the projected net imports in the year 2000 are nearly 50 million tonnes (for further discussion, see Ch. 4).

Box 3.1   *Non-food uses of food crops: bio-ethanol*
The potential of using agricultural surpluses for non-food final uses has been intensively debated in recent years, particularly in Western Europe. The possibility of absorbing large amounts of cereals or sugar in the production of bio-ethanol as a petrol-additive assumed centre stage in this debate. Interest has also been heightened by the EC decision to reduce limits on permissible levels of lead in petrol and for Member States to provide lead-free petrol by 1989. However enthusiasm for bio-ethanol programmes has been much dampened by experience in the USA and by a number of recent analyses of their economic viability. In short, bio-ethanol does not meet initial expectations and certainly it does not offer a panacea for the problems besetting the CAP.
     In the USA, the average amount of lead in all gasoline has dropped by 97 percent (from 3.5 to 0.1 gram per gallon). In 1986 the production of ethanol, in part used as an octane booster, reached 750 million gallons using 300 million bushels of grain (nearly 8 million tons) or about 3 percent of the annual US corn

production. Most of the octane gap, however, has been filled by more severe refining of crude oil and by petrochemical-based additives. The remaining market for lead substitution (and ethanol) appears therefore to be very limited and there seems to be no reason why experience in Western Europe, where the lead substitution process has only just started, should be very different.

The increase in bio-ethanol use which has occurred in the USA has only been possible with the aid of government subsidies (in the form of federal and state tax exemptions, investment incentives and import restrictions on competing products) amounting to $0.60 per gallon of ethanol. The support programme, however, does not hold up well under scrutiny. A recent study (USDA 1986b) concludes that ethanol production is a very inefficient way of absorbing agricultural surpluses and supporting farm incomes. It would be much more efficient to pay farmers directly a subsidy equal to the amount they would receive as a result of ethanol production. The study concludes that unless government subsidies scheduled to expire by the end of 1992 are extended, fuel ethanol production is likely to be terminated or sharply curtailed. A follow-up study (USDA 1988b) is more cautious in its approach and statements but leaves the conclusions of the 1986 study essentially unchanged. It states that, given likely future agricultural and energy market conditions, it might be difficult for ethanol to compete on a direct cost basis with other fuel blending agents if the fuel excise tax exemption will be discontinued. It also states that 'the non-market benefits of ethanol in meeting environmental, energy security, and agricultural goals are positive but limited, with alternatives for meeting these goals available'.

Two recently published reports are equally pessimistic regarding the economic viability of a bio-ethanol programme in the EC. The first of these states that 'at any foreseeable level of the price of oil and of the prices of agricultural feedstocks, there can be no purely economic argument in favour of bio-ethanol'. It warns against substituting (cereal) export refunds with (bio-ethanol) production refunds which 'would be seen as replacing expenditures that could be gradually reduced by expenditures for which there was an open-ended commitment' (Neville-Rolfe and Caspari 1987). The second study (CEC 1987) estimates that the total annual financial support required from the Community budget and/or Member States to encourage investment in ethanol production would be 696 to 1161 million ECU in 1990 declining to 608 to 1013 million ECU by the year 2000. This represents a level of support of 118 ECU for each of the 8.5 million tons of wheat that could be used for ethanol production in the year 2000, far in excess either of the current cereal scheme or of the proposed alternatives such as 'set-aside'. Indeed the results of a cost benefit analysis indicate that 'the encouragement of a bio-ethanol programme is not in the interests of the EC under any of the range of assumptions used'. It is against this background that the EC Commission decided (November 1987) to shelve, *sine die*, proposals for establishing a subsidised bio-ethanol industry, although it will continue to investigate the possibilities of increasing the use of agricultural commodities in the non-food sector.

4 Production, trade and self-sufficiency

Production and trade: trends

For the region as a whole the growth rate of aggregate gross agricultural production has been slowing down over the last 25 years from 2.6 percent p.a. in the 1960s and 1.9 percent in the 1970s to 1.6 percent in 1980–6 (Table 4.1). In the 1970s the slowdown was entirely due to developments in the CPE countries while the growth rates of both Western Europe and North America were maintained. The situation was reversed in the 1980s with sharp declines in the western countries being only in part compensated by better performance in Eastern Europe and the USSR.

The broad pattern is that demand factors were instrumental in the agricultural performance of the western countries while production constraints dominated that of the CPE countries. In both Western Europe and North America, production growth responded, and in general over-responded, to the rate of market expansion. For most of the 1960s it was the growth of domestic markets that underpinned production growth and there were no significant changes in aggregate agricultural self-sufficiency. In the 1970s it was the foreign trade sector that provided the scope for market expansion, mostly exports in North America and both exports and import substitution in Western Europe. Their self-sufficiency ratios increased dramatically as did the dependence of production growth on that of exports (Table 4.2). While this was nothing new for North America, the shift of Western Europe to a net exporter status for many commodities represented a radical change, with significant consequences for world agriculture.

The growing importance of exports and of import substitution is readily appreciated from the fact that 81 percent of the increment in US cereals production in the 1970s was exported and only 11 percent was absorbed by increased domestic use (the balance went to stock increases). For the EC-10, the foreign trade sector also absorbed 75 percent of the increment in production but with a fundamental difference: the bulk of it went to substitute for imports of cereals and the region became a net exporter only at the end of the decade. That this radical turnaround in the position of the EC-10 and other countries in Western Europe took place without major trade conflicts in the 1970s is to a large extent the result of the rapid market expansion in the rest of the world, USSR, Japan and the developing countries (see Alexandratos 1988, Table 3.12). When these other markets stopped growing in the aggregate during the first half of the 1980s and the momentum of production increases could not be halted, particularly in the EC-10, trade conflicts increased markedly and calls for policy reform

Table 4-1 Growth rates of gross agricultural production *

Region	Year	Commodity											
		Wheat	Coarse grains	All cereals	Oilseeds	Sugar crops	Other crops	All crops	Beef and mutton	Pig and poultry	Milk	All live-stock	Grand total
EC-north	1961-70	2.7	3.9	3.4	9.9	3.7	0.0	1.5	1.3	3.4	1.2	1.8	1.7
	1970-80	3.7	1.3	2.3	4.9	3.4	-1.4	0.7	2.3	2.4	1.9	2.1	1.6
	1980-6	5.6	0.5	3.0	15.1	-1.2	2.0	2.8	1.5	1.4	0.2	0.9	1.5
EC-south	1961-70	1.6	4.0	2.4	-0.1	6.0	1.8	1.8	4.6	8.7	1.6	4.4	2.6
	1970-80	-0.2	4.4	1.9	2.4	4.1	1.4	1.4	2.4	6.4	1.8	3.4	2.3
	1980-6	-0.7	3.8	1.8	0.6	-1.7	0.3	0.6	0.3	0.8	0.7	0.5	0.7
Scandinavia	1961-70	1.4	4.4	3.7	4.7	-0.6	-2.1	1.5	1.9	2.8	-1.4	0.1	0.5
	1970-80	0.4	1.2	1.0	4.2	4.0	-0.2	1.0	1.1	3.1	0.9	1.4	1.3
	1980-6	8.0	0.8	2.4	2.9	-0.9	2.9	2.4	0.3	0.2	0.0	0.2	0.9
Alpine countries	1961-70	3.2	4.9	4.3	3.9	4.2	-0.6	1.0	2.2	3.1	1.1	1.8	1.5
	1970-80	2.1	4.0	3.3	3.2	3.2	-1.3	0.7	2.1	3.0	0.9	1.8	1.4
	1980-6	5.6	1.7	3.0	6.9	-5.7	-2.2	0.1	1.4	1.3	1.4	1.3	0.9

Non-EC Mediterranean countries	1961-70	3.3	2.3	2.8	4.3	5.4	3.6	3.4	2.1	3.8	1.7	2.5	3.1
	1970-80	4.3	2.8	3.6	4.2	6.5	2.6	3.2	1.9	5.2	4.0	3.7	3.3
	1980-6	1.2	3.3	2.1	2.2	2.6	2.3	2.2	2.1	2.2	-0.2	1.5	2.0
TOTAL WESTERN EUROPE	1961-70	2.5	3.7	3.1	1.4	4.2	1.2	1.9	2.0	4.2	1.1	2.2	2.1
	1970-80	2.9	2.2	2.5	3.1	3.9	0.6	1.6	2.2	3.4	1.9	2.4	2.0
	1980-6	3.4	1.8	2.5	4.3	-0.9	1.4	1.9	1.3	1.3	0.3	0.8	1.3
North America	1961-70	2.4	2.9	2.8	6.0	3.0	-0.6	2.3	3.5	2.7	-0.9	1.8	2.1
	1970-80	5.0	2.9	3.5	5.5	-0.4	1.9	3.4	0.6	2.1	0.6	1.0	2.3
	1980-6	-0.6	3.0	1.8	1.5	0.0	-1.2	1.0	1.8	0.8	1.5	1.2	1.1
Eastern Europe	1961-70	5.7	1.8	3.2	5.2	1.7	1.6	2.3	5.1	2.9	2.5	3.2	2.7
	1970-80	2.3	2.7	2.5	2.8	2.0	-0.4	1.0	2.7	4.9	1.9	3.2	2.1
	1980-6	4.2	3.5	3.8	7.2	1.3	4.4	4.0	1.9	0.9	1.7	1.4	2.7
USSR	1961-70	5.2	3.1	4.4	3.9	6.8	3.5	4.0	6.2	2.2	4.0	4.3	4.1
	1970-80	0.3	1.6	0.9	-0.4	1.2	0.5	0.7	2.1	1.7	1.3	1.9	1.2
	1980-6	-1.7	5.7	1.6	0.2	2.6	2.5	2.0	2.6	3.8	2.3	2.7	2.3
ALL COUNTRIES	1961-70	3.7	3.0	3.3	4.5	4.4	1.5	2.6	3.7	3.2	1.5	2.6	2.6
	1970-80	2.4	2.5	2.5	4.2	2.1	0.7	1.9	1.6	3.0	1.5	2.0	1.9
	1980-6	0.8	3.2	2.1	2.2	0.4	1.6	1.8	1.9	1.5	1.2	1.4	1.6

* Growth rates from the gross value of production at prices of 1979/81 (for explanations see Appendix 1).

Table 4.2 Self-sufficiency ratios (percent)*

		Total	Cereals	Meat	Milk	Veg. oils and oilseeds	Sugar
Western Europe	1961/3	88	83	95	101	38	73
	1969/71	89	88	95	102	37	83
	1979/81	94	96	101	108	42	113
	1984/6	97	108	103	109	55	114
North America	1961/3	106	125	97	108	135	46
	1969/71	105	127	96	101	149	48
	1979/81	120	169	98	101	192	53
	1984/6	116	164	97	99	170	65
Eastern Europe	1961/3	97	90	102	101	72	129
	1969/71	98	93	107	100	90	94
	1979/81	97	86	110	103	89	93
	1984/6	102	96	111	106	104	98
USSR	1961/3	100	107	99	100	105	84
	1969/71	99	100	102	100	117	83
	1979/81	90	79	95	100	81	56
	1984/6	92	84	95	100	77	64

* Self-Sufficiency Ratio (SSR) equals the ratio of production over domestic consumption. It is computed from data of production and consumption (all uses) of agricultural products in primary product equivalent valued at international producer prices, the same for all countries.

multiplied as the EC-10 increased its net exports 6-fold to 20 million tonnes by 1984/6 and the USA saw its net exports fall by 26 percent to 81 million tonnes. The increased export orientation of agriculture has imparted a greater degree of volatility to aggregate demand, necessitating corresponding adaptability in the growth rate of production. Part of today's adjustment problems stems from the persistence of agricultural policies and lack of adaptability of predominantly domestic-oriented agricultures in countries which, sometimes inadvertently, became net exporters after having exhausted the scope for import substitution.

An example of differential rates of response by policy-makers and farmers in traditional and new agricultural exporting countries is provided by developments in the agricultural investment rates. During the 1970s, the period of rapid agricultural market expansion, agricultural investment rates rose in both Western Europe and North America. But reactions to the crisis of overproduction in recent years have been distinctly different. Data available for the seven major countries of EC-10 indicate that the agricultural investment rate (gross fixed capital formation in agriculture as a percent of agricultural GDP) rose spectacularly between 1970 and 1977 when the total investment rate was declining. Although it declined afterwards to 1985, it did so much less than the investment rates in other sectors. The data are as follows:

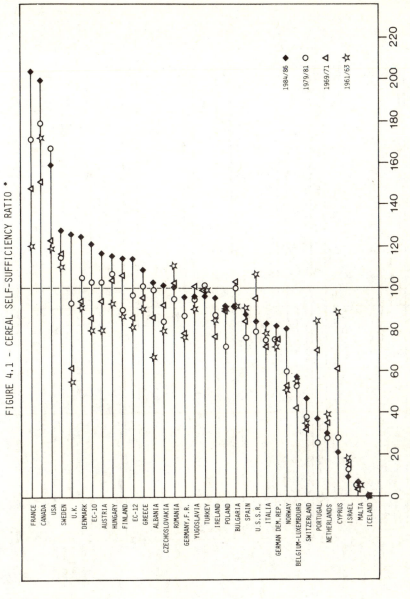

FIGURE 4.1 – CEREAL SELF-SUFFICIENCY RATIO *

* Ratio (in percent) of gross production over total domestic uses (excluding uses for stock changes).

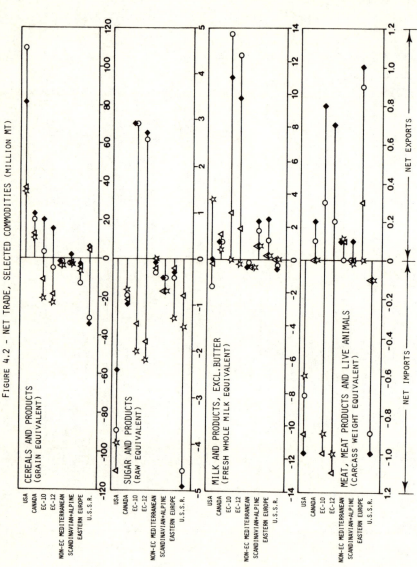

FIGURE 4.2 - NET TRADE, SELECTED COMMODITIES (MILLION MT)

EUR-7*	1970	1977	1985
Gross fixed capital formation in agriculture percent of agricultural GDP	19.4	24.0	22.9
Total gross fixed capital formation percent of total GDP	23.8	21.3	19.0

Source: Eurostat: *National Accounts ESA, Detailed Tables by Branch*, 1987, Table 12.
*   EUR-7 = EC-10 minus Greece, Ireland, Luxembourg.

The process was similar in the USA but with the significant difference that gross investments in agriculture fell dramatically over the last few years. In every year between 1981 and 1985 net capital formation was negative by a substantial amount (gross capital expenditures, including inventory changes, minus depreciation allowances — USDA 1987a, p.55). The data from the US National Accounts for gross fixed capital formation are as follows:

Gross fixed investment percent of respective GDPs	1973/75	1977/79	1983/5
In agriculture	30	34	14.6
Total economy	19.8	20.2	19.4

Source: OECD, *National Accounts, Detailed Tables 1973–85*

These differences in the investment response in the USA and the EC may be indicative of differences in both policy (the extent to which farmers are shielded by policy from market signals) and farming structures, the latter making for differing rates of response to expectations of lower farm incomes. There is a prima facie case that the system (farmers and policy makers together) underreacted in Western Europe and perhaps overreacted in the USA. In general, the high volatility of prices and farm incomes to even small shifts in demand–supply balance characteristic of farming tends to provoke overreaction.

In large measure, present problems of overcapacity are rooted in such perceptions of the period of rapid market expansion. In the 1970s, there was euphoria among actual and prospective exporters (and fears in other quarters) that the world was set on a path of rising food prices, partly based on the belief that the developing countries would continue to expand consumption faster than they could increase their production. It happened in 1973 when 'there was a widespread belief that the supply–demand imbalance had shifted permanently and that a new long-term rising trend in prices was beginning'. It was repeated in 1980 when, following drought-induced declines in yields coinciding with revival of export demand, 'there was a consensus that the situation was back on the track of a long-term trend of tight supplies' (Petit 1985). This view is also present in the scenario projections of long-term rises in food prices contained in the *Global 2000 Report to the President* (US Council on Environmental Quality and Department of State 1980).

Things turned otherwise as economic growth and that of export earnings in many developing countries came to an abrupt halt in the early 1980s and as some major importers increased their production and reduced or eliminated imports. Only two short years after the euphoria of 1980 the USA had to institute the drastic production control programme (Payment In Kind — PIK) which, helped by unfavourable weather, caused a 45 percent drop in the production of coarse grains in 1983. It all goes to demonstrate how the large impact on farm incomes, prices, stocks and budget costs produced by even small shifts in demand–supply balance tends to influence perceptions and eventually cause overreaction in policy-making.

In conclusion, the western countries amply demonstrated their capacity to increase agricultural production in response to demand increases. This was achieved with only marginal increases in total land area in agricultural use and with declining labour force. The increase in the investment rates noted earlier indicate that the capital intensity of farming increased as capital substituted for both land and labour. In parallel, a good deal of the production increases reflected yield gains coming from the application of research findings and improved management. The achievement of increased self-sufficiency and of mounting export surpluses in the main western country groups is generally considered a success story which is often cited to justify the policies pursued. However, the criterion of increased production is by itself an entirely inadequate indicator of successful development without an analysis of the related costs. In particular, the inclusion of environmental costs in the calculations may indicate that in some cases success in increasing production was bought at a high price, particularly for some products. More generally, increases in agricultural production achieved with the aid of policies involving high support costs may involve a net economic loss to society if, as usually happens, the net benefits to producers are smaller than the net costs of these policies to consumers (through higher prices) and taxpayers (Winters 1987). Recent estimates (OECD 1988:58) show that such costs amounted to nearly ECU 80 billion p.a. in 1984–6 in the EC-10 (borne mostly by consumers) as well as in the USA (borne mostly by the taxpayers).

In the CPE countries as a whole, production did not keep up with demand, the region became a significant net agricultural importer and its self-sufficiency declined. However, these developments reflected above all the situation in the USSR, while the Eastern European region as a whole, though not all countries, was quite successful in increasing production somewhat faster than domestic consumption and improving its overall self-sufficiency. Even in the USSR the longer-term (26 years, 1961–86) growth rate of production was not different from those achieved by both Western Europe and North America. Therefore one cannot really speak of production failure *per se* over the entire 25-year period, but rather of failure to match the very rapid growth of domestic consumption, as the following data show:

Growth rates % p.a., 1961–86	Consumption (all uses)	Production (gross)
USSR	2.4	1.9
Eastern Europe	1.9	2.1
Western Europe	1.4	1.9
North America	1.4	2.0

This average growth rate of the last 25 years masks, however, the fact that in the USSR production growth has been slowing down over time as follows (percent increases in gross agricultural production in successive 5-year periods; source: *Narodnoe Khozyaistvo* 1987:213):

From 1961–65 to 66–70	+ 21 percent
From 1966–70 to 71–75	+ 13 percent
From 1971–75 to 76–80	+ 9 percent
From 1976–80 to 81–85	+ 5 percent

Inadequate agricultural investment is not a satisfactory explanation for this continuous fall in the agricultural growth rate, since the sector absorbs about one-quarter of all material sphere investment while it contributes only some 12 percent of net material product (both shares at 1976 prices); alternative estimates at prices of more recent years put agriculture's share nearer 20 percent (UN/ECE 1988:225). This reflects in part the faster growth of agricultural producer prices compared to those in the other material sphere sectors. Such heavy investment allocations failed to produce the 'normal' developmental pattern of rising labour productivity (net output per worker) which 'declined in absolute terms in 1976–80 and again in the first half of the 1980s' (UN/ECE 1988: 225). It is argued in Chapter 6 that, although the harsher ecological conditions prevailing in USSR agriculture compared with other countries would justify higher use of capital per unit of output[1], too much was being invested in primary agriculture *in relation to the results obtained*, rather than too little. More investment may not have increased the growth rate of production unless the more fundamental causes of declining productivity of capital had also been tackled. For example,

---

[1] Unfavourable agro-climatic conditions are often cited as an important factor in the development of USSR agriculture (for elaboration see comparisons with the USA by former Agriculture Minister V.V. Matskevitch in Ogonèk, January 1973, cited in UN/ECE 1988:224). This should not be exaggerated, however, since the USSR is also endowed with plentiful good quality agricultural resources; it is 'the country in which 80 percent of the "black earth" soil in the world is concentrated, the best soil, with enormous richness, fertility and the most varied of conditions' (Aganbegyan 1988:180).

there was a predominance of administrative procedures, while economic factors seemed to have had a subsidiary role as criteria in key decisions affecting production, marketing and, more generally, the use of resources. As a result, there has been a proliferation of the number of cooperative and, even more, state farms plagued with deficits and debts. A review of the relevant literature on the subject suggests that after the early 1970s and until quite recently more than average incidence of unfavourable weather coincided with the period when the effects of these inappropriate policies came to a head.

Up to quite recently, there has been an apparent excessive reliance on investment allocations as the single most important instrument of agricultural policy, including lack of balance between investment allocations to primary farming (excessive) and those in the allied sectors (transport, processing, storage, input industries), with the result that post-harvest losses have been very high. Visible effects of these policies have been the failure to mitigate the wide year-to-year fluctuations in yields and production (half of which may be ascribed to instability in the supplies of inputs, Desai 1986), the persistence of inadequate and unbalanced feed supplies in relation to the total livestock population and outright long-term declines in yields (e.g. of sugar beet) due to the non-adoption of appropriate technologies. Under the policy reforms undertaken in recent years more balance is being introduced in investment allocations. For example, in the twelfth 5-year plan the agricultural machine-builders 'were allotted almost six billion additional rubles in capital investment at agriculture's expense' (Guzhvin 1987:49). Aganbegyan (1988:181) also states that 'We have invested greatly in storage and more thorough processing of agricultural produce' (for further discussion of production policies, see Ch. 6).

Production and trade: projections

The preceding discussion has made it clear that production levels in the majority of the countries of this study are greatly influenced by policy interventions. Such policies work through the trade linkages to influence production also in other countries which follow more market-oriented policies. Projecting production for the major country groups individually would, therefore, require assumptions as to future policies. Such assumptions are impossible to make in the current climate of ongoing negotiations aimed at multilaterally coordinated policy reforms.

Following the approach of the study *World Agriculture: Toward 2000*, only a rough extrapolation of production trends is attempted here for country groups of both study and non-study developed countries. These are subsequently adjusted only for the CPEs but not for individual groups of the developed market economies. The resulting imbalances between the supply and demand for the major commodities are subsequently confronted with the net trade positions of the developing countries as projected in the above-mentioned study. These quantifications serve to provide a rough quantitative framework for the subsequent discussion of policy issues.

For livestock products and for crops for which the area and yield data are incomplete production is extrapolated directly as a function of time, using formulas which attenuate the historical trends of the last 10 or 15 years when such trends imply explosive and generally infeasible future growth patterns. For the other crops these same extrapolations are performed separately for area and yield. By their nature these extrapolations indicate future production levels which will not materialize if they are not compatible with market balance conditions and/or policy stances. This being so, not much effort need be devoted to devising a 'perfect' formula for measuring and extrapolating trends, or for selecting the most appropriate historical period.

Livestock products

The region as a whole is a net exporter of meat and milk. For meat, the net exporter status of the region is a quite recent phenomenon, due mostly to import substitution and export expansion in most countries of both Western and Eastern Europe. Net exports to the rest of the world at 195 thousand tonnes in 1984/6 are a very small proportion of total production — about 0.2 percent. It is unlikely that the rest of the world will become a major net importer of meat from the countries included in this study. Therefore net trade with the rest of the world should remain a minor factor in the growth of the livestock sector of the region as a whole, though exports and imports within the region and even outside it will continue to be very important for individual countries.

The projections of the *AT 2000* study indicate that the developing countries as a whole would continue to be net importers of meat (mostly sheep meat and poultry meat) at just over present levels. The net exports of the CPEs have been fluctuating widely and exhibited no consistent trend in the historical period. Therefore the trend production projections of the CPEs are adjusted to maintain net exports at around present levels. The trend extrapolations of production in the developed market economies (DMEs), even if considerably attenuated, would tend to produce an exportable surplus well in excess of the net import requirements of the rest of the world (Table 4.3). This imbalance gives an indication of the magnitude of required adjustment in the growth rate of production, from 1.5 to just under 1.0 percent p.a. for 1984/6–2000. This 'permissible' growth rate of production reflects above all the growth rate of domestic demand for meat implied by the projected per caput levels (see Table 3.2).

For milk, the dependence of production on exports is considerable for many northern countries of Western Europe as well as for North America and some countries of Eastern Europe. The region, however, includes also major importers; the USSR and the countries of Southern Europe. The region as a whole is a net exporter to the rest of the world of 14.5 million tonnes of milk and products (fresh whole milk equivalent, not counting butter) which is 4.0 percent of gross production, although this percentage is 18 percent in the EC-North (average for 1984/6). Thus, unlike meat, net trade to the rest of the world is an important factor in production growth. The *AT 2000* projections for the developing countries indicate that their net

Table 4.3 Meat and milk production and trade

	Production growth rates 1971-86 1984/6–2000 percent p.a.		Net trade 1984/6 2000 '000 tonnes	
a. Meat (carcass weight, excl. offals)				
CPEs	2.0	1.5 trend	50	1,500 trend
		1.3 adjusted		80 adjusted
Developing countries	4.7	3.7 adjusted	−630	−800 adjusted
Sub-total			−580	−720 adjusted
DMEs*	2.1	1.5 trend	1,140	7,720 trend
World imbalance			+560[†]	+7,000
b. Milk (fresh equivalent, excl. butter)				
CPEs	1.1	1.0 trend	2,060	−1,000 trend
		1.3 adjusted		+2,200 adjusted
Developing countries	3.5	3.4 adjusted	−20,100	−27,500 adjusted
Sub-total			−18,040	−25,300 adjusted
DMEs*	1.5	1.3 trend	18,110	39,000 trend
World imbalance			+70[†]	+13,700

* Developed market economies: Western Europe and North America as defined in this study plus Japan, Australia, New Zealand, and the Republic of South Africa.
† Statistical discrepancy.

import requirements would continue to grow, though at a much slower rate than in the past. Such import growth would depend, among other things, on the continued availability of a substantial part of total imports on concessional terms, including in the form of food aid. Under these assumptions the net imports of the developing countries could amount to some 27.5 million tonnes by 2000, a growth rate of 2.1 percent p.a. from the average 1984/6 compared with 7.8 percent in the preceding 15 years. The dependence of these imports on concessional sales, however, introduces a high degree of uncertainty into these projections. Current policies, particularly in Europe, aiming at restraining the growth of production and in encouraging domestic use, particularly of skimmed milk powder in the feed sector, may result in concessional exports being more limited in the future compared with the past. It is therefore possible that developing country consumption and net imports will not grow as fast as implied by these projections.[1]

[1] The following quotation from *World Agriculture: Toward 2000* (Alexandratos 1988:108) elaborates on the projected milk import requirements of the developing countries as follows: 'The growing net deficit of the 93 developing countries (outside China) of 23.7 million tonnes in year 2000 (of which 19 million tonnes in the middle-income countries) is based on projected per caput consumption growing to 51 kg by 2000 compared with 46 kg in 1983/5 and 38 kg in 1969/71. In parallel their production is projected to grow at 3.1 percent p.a., roughly the same rate as for 1970–85. Their net imports grew very fast in the past 15 years, 7.8 percent p.a. The above projected deficit means that they would grow at only 1.9

The trend extrapolations of production in the CPEs imply a deficit in the year 2000 for the USSR. They are adjusted upwards to maintain net exports of the CPE region as a whole at around current levels. With this adjustment, the net import requirements of the non-DME world would be around 25 million tonnes, some 40 percent above present levels. A conservative trend extrapolation of production in the DMEs would lead to an exportable surplus in excess of these import requirements (Table 4.3). Therefore, production growth must be restrained further to an average of 0.9 percent p.a. This would probably be achieved by a continuation of existing production controls in the main producing countries, the effects of which are not fully reflected in the continuation of trends of the past 10 or 15 years.

Major crops

The trend extrapolations of crop production are derived, for each crop in each country group, from separate extrapolations of land area, total and by crop, and of yields, as described in the Appendix. This method is used for all crops of the Study except those for which consistent time series data for area and yields are not available for many countries (mostly tree crops). For these latter crops trend extrapolations are derived directly for total production. The extrapolations of aggregate harvested area reproduce the historical trends in somewhat attenuated form. For the study countries as a whole, the total area under the crops examined increases by 10 million ha to 392 million ha mostly because the extrapolations pick up the historical trends towards increases in North America (Canada) and, to a smaller extent, in some other country groups (EC-North, non-EC Mediterranean) notwithstanding declines or near constancy in the other country groups.

The distribution of this total area among the major groups of crops also reproduces historical trends in attenuated form. Thus in both EC-North and North America the share of total area going to wheat continues to increase and that of coarse grains to decline. The opposite is true for the Southern countries (EC-South and non-EC Mediterranean) as well as for the USSR. These divergent trends have not been independent of each other. For example, the changing cropping patterns in EC-North have been linked to those of EC-South; and the expansion of wheat area in both EC-North and North America were partly related to the contraction in the USSR. The most notable historical trend is the rapid increase in the share of total area

percent p.a. in the future. The projected import requirements rise to around 27 million tonnes if an allowance is made for China and the other developing countries which currently import 2.2 million tonnes. These projections are subject to more than the "usual" degree of uncertainty since a good proportion of developing country imports in the past were concessional, including food aid, which in 1983/5 accounted for 28 percent of all imports of skimmed milk powder, the main dairy commodity imported. Food aid volumes and prices of commercial imports are likely to be affected by policies to curtail growth in milk production in Western Europe and the USA. Some other studies also project milk import requirements of the developing countries. The IIASA study cited earlier (Parikh *et al.* 1988) projects such imports at 23.5 million tonnes (reference scenario) while a trend projections study of the International Food Policy Research Institute (IFPRI, Sarma and Yeung 1985) has much higher net import requirements, 61.9 million tons for the 90 developing countries of the 1981 edition of *Agriculture: Toward 2000'*.

devoted to oilcrops everywhere except in the USSR. This reflected above all
the rapid growth in the use of oilcakes in the feed sector and the support
policies in many countries. It contributed, however, to the creation of
oversupply conditions in the oils and fats sector at the global level. These
trends in the cropping patterns are not necessarily sustainable in the future.
The extent and directions in which they may be modified for world balance
are discussed later.

Total projected trend production is derived by multiplying projected land
under each crop with extrapolated yields. The latter are shown in Table 4.4.
By and large these extrapolations imply continued strong growth in
productivity, with the notable exception of the USSR. In examining these
extrapolations two questions may be raised: first, whether the projected
increases of average yields to very high levels by today's standards, e.g. 8.4
tonnes/ha for wheat, 9–10 tonnes/ha for maize or over 6 tonnes/ha for barley
are realistic from an agronomic/biological standpoint. The answer here is
yes, at least according to the great majority of experts. Reported record
yields of wheat and maize are 14.5 and 19.3 tonnes/ha, respectively
(Coffman 1983). It is also suggested that some technological advances,
already in the pipeline,

may dwarf those of the Green Revolution which came into prominence in the 1970s.
In a dramatic way, the differences between the new super-productive barley hybrids
and their predecessors have been likened to the contrast between the 'Train à
Grande Vitesse' (TGV) and a funicular railway. (Ollerenshaw and Last 1988)

The second question is whether such yield increases are environmentally
sound and sustainable given the high rates of agrochemical applications that
would be required. There is no easy answer to this question, but the current
problems associated with high use of nitrogenous fertilizers are telling (see
Ch. 9).

Related to the above is the issue whether so highly intensive farming is
really needed, if part of the output so produced is surplus to regional and
even global requirements at prevailing prices. This is an important question
in the current debate towards policy reform, i.e. whether lower production
growth is to be achieved through more intensive use of constant or reduced
land area or through more extensive use of more land. Economists would
respond that structural surpluses (those which do not arise from fortuitous
causes, e.g. good weather) are the result of inappropriate farm support
policies, including those which stimulate production in the pursuit of non-
production objectives, e.g. farm incomes, regional balance, etc. From this
standpoint, the attainment of improved market balance would be facilitated
by increased agricultural productivity leading to more efficient and viable
farms, provided all environmental costs are internalized. This is because
more efficient farms would reduce the need for farm support (for further
discussion, see Ch. 5).

A glance at Table 4.4 brings out the wide differentials in prevailing and
trend-projected yields in the different country groups. By themselves these
differentials do not tell much, since the lower yields associated with
extensive farming in land-abundant regions are part of farming systems
which are as efficient economically as those of the regions with very high

yields, and sometimes more so. Thus, wheat production costs in particular regions of Australia and Canada with low yields compare favourably with those of some intensively farmed areas of Europe, which achieve yields of three to seven times as high (Murphy 1988). There is evidence, however, that the productivity gains achieved behind the high support and protection regime of the EC have resulted in the emergence of a significant part of the region's cereals sector as a highly efficient agricultural system capable of withstanding international competition even at the much lower world market prices (Stanton 1986; see also Ch. 12). It must be noted, however, that international comparisons of production costs are fraught with difficulties because of fluctuating exchange rates and other problems, e.g. how to account for land costs when land rents are themselves a function of profitability of production. In some cases the persistence of low yields are a good approximate indicator of failure to develop an efficient agriculture, e.g. when increased output and improved self-sufficiency are highly valued both economically and politically, e.g. in the USSR. The yield increase requirements in such cases are discussed below.

Cereals

The trend projections for the CPE region reflect essentially the continuation of the historical declines in the growth rate of production in the USSR. If these projections materialize they would lead to further increases in the region's net import requirements to around 60 million tonnes. This is above the highest net imports on record of 55 million tonnes in the 1984/5 crop year. However, an ever-expanding USSR deficit is not a realistic outcome in view of the serious efforts under way towards policy reform precisely to halt and reverse these trends as well as in view of the increasing balance of payments constraints and of the encouraging production levels achieved in recent years. Projections of other studies indicate a wide array of cereals deficits for the CPE region for the year 2000 ranging from 20 million tonnes (International Wheat Council 1983, 1987; their CPE region includes Cuba, though, which in 1984/6 was a net importer of 2.2 million tonnes of cereals) to 31 million tonnes (IIASA, Parikh *et al.* 1988:214, reference scenario projections) and 88 million tonnes (World Bank 1986).

A recent article in the OECD Observer (Kuba 1988) is generally optimistic on the prospects of the USSR achieving full self-sufficiency in cereals by the mid-1990s. Similar optimism is also present in the July 1988 issue of the International Wheat Council's projections, where it is stated that 'It is assumed that the new agricultural policies being adopted in the USSR will, through increased production and a more efficient use of grain, result in that country becoming practically self-sufficient by the year 2000' (IWC 1988:32). In the same direction, the 1989 edition of the World Bank's projections have lowered the projected cereals deficit for the CPE region to 69 million tonnes. This reduction reflects entirely the better production prospects assessed for the USSR (225 million tonnes, practically identical to the production projections of this study) and somewhat lower levels of projected consumption, resulting in a net projected-year-2000 deficit of

Table 4.4 Trend projections of yields (tonnes/ha)

		Wheat	Maize	Barley	Other coarse grains	Soya beans	Sunflower seed	Rapeseed	Sugar beet
EC-north	74/6	4.3	4.3	3.7	3.3	*	1.5	2.1	40
	84/6	6.3	6.3	5.1	4.3	2.1	2.1	2.9	52
	2000	8.4	8.1	6.2	5.0	...	2.5	3.8	63
EC-south	74/6	2.1	4.2	1.9	1.3	1.6	0.6	*	41
	84/6	2.6	6.5	2.4	1.5	3.3	1.1	1.7	45
	2000	3.1	9.4	2.8	1.6	3.5	1.4	...	53
Scandinavia	74/6	4.1		3.1			2.8	2.0	36
	84/6	4.6		3.4	3.5			2.0	37
	2000	5.6		3.9	4.1			2.0	37
Alpine countries	74/6	4.1	6.1	3.8	3.3			2.3	48
	84/6	4.9	7.9	4.5	3.9		*	2.7	54
	2000	6.1	10.0	5.5	4.7	...	*	3.3	66
Non-EC Mediterranean countries	74/6	1.8	3.4	1.7	1.3	1.7	1.3	2.0	37
	84/6	2.2	4.5	2.1	1.6	2.1	1.4	2.2	35
	2000	2.7	6.3	2.7	2.1	2.7	1.4	2.7	35

North America	74/6	1.9	5.2	2.2	2.3	1.8	1.1	1.0	43
	84/6	2.3	7.1	2.7	3.1	2.1	1.3	1.3	46
	2000	2.7	8.7	3.3	3.2	2.4	1.5	1.6	46
Eastern Europe	74/6	3.2	3.5	3.3	2.5	1.5	1.4	2.2	28
	84/6	3.9	5.4	4.0	2.8	1.3	1.9	2.3	32
	2000	5.0	7.6	4.6	3.1	1.2	2.3	2.8	37
USSR	74/6	1.4	3.0	1.6	1.3	0.7	1.3	1.0	22
	84/6	1.6	3.2	1.6	1.5	0.7	1.3	0.6	24
	2000	1.6	3.9	1.6	1.5	0.7	1.2	0.4	23
Total	74/6	1.9	4.6	2.2	1.9	1.7	1.2	1.6	30
	84/6	2.4	6.3	2.5	2.2	2.1	1.4	1.8	34
	2000	3.1	7.9	2.9	2.4	2.4	1.5	2.3	36

* Only small quantities produced.

Note: A recent study prepared for the EC Commission projects yields in the EC-12 to grow between the mid-1980s and the year 2005 at the following rates (percent p.a.): soft wheat 2.0, barley 1.5, maize 1.5, sugar beet 1.5, rapeseed 2.0, sunflower seed 2.4, soyabeans 2.0. These yield growth rates are based on the authors' judgement, assuming that EC agricultural policy continued along present lines, which more or less ensure that agricultural prices are cut in line with the rate of productivity improvement in agriculture. The Study then considers that biotechnology applications could increase average yields further, so by the year 2005 crop production could be some 5 percent above that implied by the above 'secular' trends in yields (CEC 1989d: 49, 61–2).

some 50 million tonnes (World Bank 1989). A more recent assessment is that USSR grain imports should generally total 25–40 million tonnes a year through 1995 (Johnson 1989).

The assessment of the CPE region as a whole made more than a year ago in the context of the *AT 2000* study was for a net cereals deficit by year 2000 of around 35 million tonnes. Given the past experience with rapidly increasing deficits, this projected deficit reflected very guarded optimism concerning the adoption of required policy reform. One can be somewhat more optimistic today when efforts at policy reform have received wider acceptance. One way of defining an 'optimistic' cereals deficit outcome for the USSR would be to assume a level of just under 30 million tonnes which is equal to the average of the last three years 1986–8. It is optimistic in the sense that these imports were necessary to meet domestic demand[1] even in years in which production was well above the annual average of the first half of the 1980s.

This approach may, however, overstate the case that the primary role of imports in the context of the USSR policy-making has been to make up for actual or anticipated shortfalls in domestic production in face of the need to maintain and increase consumption levels of livestock products (see, for example, Ennew 1987). Recent econometric work (Borsody 1987) lends strong support to the hypothesis that more conventional economic factors have been at play in determining the USSR's cereal imports. These are the significant gains in the country's terms of trade and foreign exchange availabilities following the oil and gold price rises from the early 1970s to the early 1980s, in combination with the decline in the world prices of cereals. By contrast, the domestic production of cereals is shown to have been only weakly related to the decision to import. It is not possible to forecast future developments in these economic factors. Some price forecasts indicate continued weakness in the real (constant dollars) prices of cereals and gold while those of oil could recover from the depressed levels of 1988, although in 2000 they would still be around one half of their high average level of the early 1980s. Moreover, progressive integration of the country into the world economy could diversify the sources of foreign exchange earnings and introduce new elements in the economic evaluation of food imports and the self-sufficiency objective. Additionally, the objective of maintaining and

[1] It is assumed that the USSR will continue to prefer to import feedstuffs rather than livestock products. This makes good economic sense assuming productivity in feed use can be improved substantially. Alternatively, imports of livestock products could be considered instead. However, the comparative economics of these options would depend significantly on, among other things, developments in world market prices of cereals, oilseeds and livestock products, including prices and conditions as determined by subsidized sales of major exporting countries. They would also depend on the extent to which weaknesses in the domestic processing, transport and distribution sectors are overcome. If they are not, imports of consumer-ready products (poultry, butter) have certain advantages as a short-term policy response (see IWC 1989).

improving the consumption levels of livestock products may gain in relative importance in the context of the new policy environment. Consequently, a closer association between domestic production and imports may prevail in the future compared with the past.

This latter hypothesis is supported by more recent research (Greenfield 1989). It shows some significant differences between the 1970s and the 1980s as to the extent to which variations in grain (wheat and coarse grains) production in the USSR were absorbed through changes in the country's net imports, consumption and stocks. In the 1970s the impact of production variations was reflected more on consumption and less on net imports. The opposite was the case in the 1980s, as the following data show:

USSR: Percentage of grain production variation reflected in:

Crop years	Net imports	Stocks	Consumption
1972/73–1982/83	27	39	35
1979/80–1988/89	44	36	20

In addition, there was a statistically significant correlation in the 1970s (but not in the 1980s) between changes in net grain imports and those in the 'real' price of grain in the international markets (real = price of grain deflated by price of petroleum).

For the containment of the cereal deficits at the levels indicated in the preceding paragraphs, USSR production in an average year should increase to 227 million tonnes by 2000, implying a growth rate of 1.0 percent p.a. if measured from the 200 million tonne level of 1987, or 1.2 percent if measured from the three-year average 1985/7 of 194 million tonnes. These are certainly not spectacular growth rates, but they would represent a radical departure from past trends. In the past 20 years (1966–86) there was not one single 15 year period (1966–81, 1967–82, ..., 1971–86) in which the production growth rate of cereals was statistically different from zero. At the same time, it would 'only' take maintenance of present area under cereals (111 million ha in 1984/6) and an increase in average cereal yields from 1.63 to 2.05 tonnes/ha to reach a projected production of 227 million tonnes. Table 4.4 shows that cereal yields in all other country groups of this study exceed these numbers, often by a large margin (see, however, earlier discussion concerning the generally harsher agro-climatic conditions of the USSR). If policy reform were to proceed seriously there is no reason why these 'required' yields could not be achieved in the USSR. As a minimum, such policy reform could contribute to reduce the very wide yield variability by increasing the reliability in the supply of inputs.

The net cereals deficit of the CPE region projected as above together with that of the developing countries as projected in the *AT 2000* study (see Box 4.1 and Table 4.6) imply that the net import requirements of the non-DME world would be around 136 million tonnes compared with 105 million tonnes in 1984/6 (Table 4.5). This would provide only limited scope for further

Table 4.5 Cereals: production and trade

	Production growth rates		Net trade	
	1971–86	1984/6–2000	1984/86	2000
		percent p.a.	million tonnes	
CPEs	0.7	0.4 trend	−39.0	−61 trend
		1.3 adjusted		−26 adjusted
Developing countries	3.4	2.7 adjusted	−65.6	−110 adjusted
Sub-total			−104.6	−136 adjusted
DMEs	2.4	1.4	110.6	236
World imbalance			+6.0	+100

BOX 4.1 * *Factors in the growth of cereals deficits of the developing Countries*
The argument of the required adjustment in the growth rate of cereals production assumes as given the projected net import requirements of the developing countries (Table 4.6). These projections may appear to be on the conservative side in the light of past experience. They are more so if viewed in the context of the widely held belief that the developing countries are embarked firmly on a path of increasing consumption of livestock products which will continue to fuel import demand for feedgrains even if their own cereals sector were to grow reasonably fast. Empirical evidence to support this argument is mixed. It is true that in countries with rapidly rising incomes the demand for livestock products is strong and imports of feedgrains grow rapidly. It is, however, equally true that in many developing countries the rapid growth in cereal imports was caused by the slow growth in their production rather than by acceleration in their (food and feed) consumption of cereals. In these countries there is a strong import replacement effect when production growth improves. This is not to deny that in countries in which improved agricultural growth is an integral part of, and stimulates, sustained economic growth, the income effect will tend to overshadow the import replacement effect (see for example Houck 1988). However, the net effect will depend on how many developing countries may fall in this category in the next 15 years. There are major cereal importers in both classes of countries. On balance, therefore, what will happen to the growth rate of net import requirements of the developing countries will depend on whether the income effect or the import displacement effect will dominate in the future. Optimism concerning the prevalence of the income effect in the future is not uncommon. For example, Hooke (1989) projects for the developing countries of Asia food demand to grow over the longer term (to the year 2050) at 4.5 percent p.a. and production at 2.6 percent, implying

spectacular growth in Asia's food import requirements. The FAO study *World Agriculture: Toward 2000* concludes that given the less favourable income growth prospects in many middle-income countries and the increasing foreign exchange constraints compared with the past, it is reasonable to expect that acceleration of consumption spilling over into rapid growth of imports will be less of a factor in the future compared with the past. The import scene may, therefore, be determined more by the extent to which importing countries are successful in maintaining or improving their production performance.

Other projection studies indicate cereal deficits of the developing countries which, in most cases, are above those projected in *World Agriculture: Toward 2000*, as follows (million tonnes, deficits derived as difference between projected production and consumption):

Developing Countries, Cereal Deficits in 2000

	Total	China and India	Latin America	All Others
FAO (Alexandratos 1988)	-112	-11	1	-102
World Bank (1989)	-206	-62	-1	-143
International Wheat Council (1988)	-136	-28	-6	-102
IIASA (Parikh *et al.* 1988) [1]	-146	-15		
IFPRI (Paulino 1986) [2]	-80			
FAPRI (1988) [3]	-104			

[1] Deficit shown is for the study's reference scenario.
[2] Deficit shown is for the 90 countries of the 1981 edition of AT 2000, which account for 93 out of the 112 million tonnes shown above.
[3] Deficit shown is for 1996/7 and excludes rice.

The above numbers are not fully comparable with each other because of some differences in the countries classified as developing. These differences in country coverage are not, however, very significant for explaining the discrepancies in projected deficits. Controlling for differences in country definitions, the difference in the projected deficits between the FAO and the World Bank is almost entirely due to the latter's lower production projection for the developing countries, while projected demand is nearly the same in the two studies. A fuller explanation of differences cannot be made without a detailed comparative study of country level results as well as of the methodology and assumptions, a task which is not possible from the published material. The regional differences that could be identified are shown above. The deficits, even when not very different from each other, are the results of projected production and consumption which can differ widely among studies, including in commodity composition (wheat, rice, coarse grains) and in the distribution of cereal consumption between food and feed.

* Adapted from Alexandratos (1988) and updated.

expansion of net exports of the DMEs which played such an important role in their maintenance of high production growth rates in the past. Even the conservative projection of their production trends implied by the land and yield extrapolations would lead to an exportable surplus which would exceed the above import requirements by some 100 million tonnes.

If the whole adjustment for world balance were to be borne by the DME group of countries, their growth rate of cereals production between 1984/6 and 2000 should not exceed 0.5 percent p.a. on the average. However, measuring growth from a three-year average of 1984/6 understates the adjustment that is currently taking place and has led to 1987 cereals output of the DMEs being 9 percent below that of this three-year average. The production growth required for world balance would be 1.3 percent p.a. if measured from the reduced 1987 production. Although this may look encouraging, account must be taken of the fact that the 1987 'adjustment' was borne predominantly by the USA (12 percent decline in production), Canada (10 percent), Oceania (17 percent) and the Scandinavian countries (14 percent) while output in the EC-12 remained constant. These 1987 declines were policy-induced in the USA when land idled under government programmes increased from 44.8 million acres in 1986 to 69.3 million acres in 1987 (USDA 1987c). They are likely to be reversed as soon as the stocks and market outlook improve (see Ch. 1). Therefore, a long-term production growth rate of under 1 percent p.a. may be a better indicator of pressure to adjust than the 1.3 percent rate measured from the 1987 levels.

Sugar

The trend extrapolations of sugar beet production in the CPE countries imply negative growth of production. This is the effect of a continuation of declines in the USSR and slow growth in Eastern Europe. The region's deficit would increase to 7.4 million tons (raw sugar equivalent), all of it in the USSR. However, the trends of the USSR paint too pessimistic a picture in view of efforts undertaken in recent years to improve sugar beet production and the results achieved (promotion of intensive technology, adopted in 70 percent of the beet area by 1985; it is estimated that production in both 1987 and 1988 was consistently above trend, at 9.8 and 9.6 million tonnes, respectively). Assuming net imports would remain at around present levels of 5 million tonnes, USSR production of raw sugar would need to grow over the next 15 years at 1.0 percent p.a., compared with virtually no growth in the 15 years to 1986. This historical period was characterized by wide fluctuations, from 9.3 million tonnes in 1970 to 6.2

Table 4.6 Net cereals balances of the developing countries by major importers and exporters * (million tonnes)

	1969/71	1979/81	1984/6	1987	1988	2000
Developing net						
importers	−34.4	−88.5	−93.5	−103.3	−106.9	−157
Oil exporters †	−5.0	−26.8	−33.6	−34.1	−33.7	−56
Saudi Arabia	−0.5	−3.1	−7.5	−7.3	−3.3	
Algeria	−0.5	−3.0	−4.6	−3.8	−6.1	
Mexico	−0.2	−5.9	−4.4	−4.7	−6.0	
Iran	−0.5	−2.7	−4.1	−5.3	−4.4	
Iraq	−0.4	−2.7	−3.9	−4.2	−4.3	
Venezuela	−1.0	−2.5	−2.6	−2.3	−3.4	
Nigeria	−0.4	−2.1	−1.8	−0.8	−0.9	
Indonesia	−0.9	−2.6	−1.5	−2.0	−1.7	
Libya	−0.3	−0.7	−1.3	−1.4	−1.4	
Others‡	−0.7	−1.5	−1.9	−2.3	−2.2	
Other net importers	−29.4	−61.7	−59.9	−69.2	−73.2	−101
Egypt	−1.1	−5.9	−8.3	−9.1	−8.2	
Korea, Rep.	−2.5	−5.7	−6.8	−8.7	−9.3	
China (incl. Taiwan						
Province)	−3.8	−15.9	−6.6	−16.3	−16.5	
Brazil	−1.0	−6.3	−5.5	−4.2	−1.8	
Cuba	−1.2	−2.1	−2.2	−2.4	−2.5	
Morocco	−0.3	−2.0	−2.2	−2.2	−1.4	
Malaysia	−0.9	−1.6	−2.2	−2.2	−2.5	
Bangladesh	−1.3	−1.4	−1.9	−1.8	−3.0	
Peru	−0.7	−1.3	−1.4	−1.6	−1.7	
Syria	−0.4	−0.7	−1.4	−1.3	−1.0	
Philippines	−0.8	−0.9	−1.4	−1.0	−1.5	
Tunisia	−0.4	−0.9	−1.0	−1.2	−2.1	
India	−3.6	0.4	−0.3	0.7	−2.6	
Others‡	−11.4	−17.4	−18.7	−17.9	−19.1	
Developing net						
exporters	14.0	21.4	26.5	17.6	21.1	45
Argentina	9.4	14.4	17.2	9.4	10.0	
Thailand	2.9	5.2	7.9	6.1	6.6	
Others‡	1.7	8.8	1.4	2.1	4.5	
All developing						
countries‡	−20.4	−67.1	−67.0	−85.7	−85.8	−112

Source: Alexandratos (1988), Table 3.12, with historical data revised and updated to 1988.
*   A minus sign denotes net imports. All quantities include rice in milled terms. For the purposes of this table, Turkey and Cyprus are included in the developing countries as in the original source given below.
†   IMF classification of countries (18) in which fuel exports accounted for more than 50% of total exports in 1984/6 (IMF, *World Economic Outlook*, 1989).
‡   Including developing countries not included in the 94 study countries of *World Agriculture: Toward 2000*. Countries listed individually had net imports of 1 million tonnes or more in 1984/6, except for India.

million tonnes in 1981 to the three year average of 8.5 million tonnes in 1984/6. Achieving an average growth rate of 1 percent p.a. in 1984/6–2000 would require departure from past trends of stagnant or declining sugar beet yields, to make them increase from 24 to 31 tonnes/ha. The achievement of yields of around 26.0 tonnes/ha in both 1987 and 1988 is encouraging in this respect. This would still leave year 2000 yields somewhat below those currently achieved in the east European countries and well below those in all other country groups of the study.

The net exports of the developing countries as projected in the FAO study *World Agriculture: Toward 2000* are around 7.0 million tonnes compared with some 6.0 million tonnes in 1984/6 and some 7.0 million tonnes in 1979/81. The general assumption underlying the projections of the developing country exports for commodities like sugar, which compete with production in the importing developed countries, is that current efforts at policy reform, particularly in these importing countries, will at least succeed in halting the trend towards decline in these net exports. Given, however, the great uncertainties surrounding the prospects for policy reform, a wide range of alternative trade outcomes is possible.

For the same reason, production trends in the DME countries are very difficult to project. In recent years production in Western Europe has been restrained through quotas and encouraged in North America through import restrictions in the USA. Net imports of the DME group of around 2 million tonnes in 2000 (the net balance of the developing countries and the CPEs) would be feasible if their production growth rate were limited to around 0.6 percent p.a. This is higher than their production growth rate of the 1980s (0.4 percent p.a. in 1980–7) but well below that of the 15-year period 1972–87 which was 2.1 percent.

Oilseeds

In the CPE countries the trend extrapolations of area and yields of oilseeds (including cotton) lead to growth rates of production of 1.4 percent p.a. (in terms of oil equivalent). This is well above the growth rates achieved in any 15 year period from the mid-1960s to the present, although it is lower than the growth rate of more recent years (2.3 percent in 1980–8). The trend production in 2000 implies continued growth in the net import requirements of the CPE region as a whole, with self-sufficiency declining further to 77 percent from 84 percent in 1984/0 (it was 110 percent in 1969/71). All these data and projections are in oil equivalent and the net trade and self-sufficiency estimates reflect the comparison of domestic demand for vegetable oils and the oil equivalent of domestic oilseed production. The self-sufficiency situation is different in terms of the oilmeals. The oilmeal equivalent of domestic oilseeds would cover only 53 percent of the feed requirements of some 15 million tonnes projected in the preceding section. This rate of self-sufficiency is roughly equal to the present one, but it must be remembered that the demand for feed was projected in very conservative terms and assumed, among other things, more emphasis on the development of alternative indigenous sources of proteins.

The projections of the FAO Study *World Agriculture: Toward 2000* of exports from the developing countries are subject to the same policy qualifications as for sugar, i.e. much depends on the agricultural support and protection policies of some major importing DMEs. They indicate 3.1 million tonnes as net exports in 2000. The projected net import requirements of the CPEs would offset over one half of this exportable surplus leaving the balance to be absorbed as net imports of the DMEs. This is about the same level of net imports as in recent years.

The trend extrapolations of the DMEs point, however, to continuous strong growth in production. If they materialized, the group would turn from net importer to net exporter of oilseeds and oils (oil equivalent): the net export surplus of North America would continue to grow while the net import requirements of Western Europe would continue to decline. These implications of production trend continuation would reflect the radical shifts which occurred in the 1980s and are still under way, with Western Europe increasing its share in combined (Western European and North American) production of the main oilseeds (soybeans, sunflower seed, rapeseed) from 6 percent in 1979/81 to 12 percent in 1984/6 and 18 percent in 1987. Measured in oil equivalent, however, and taking into account other oilseeds and Western Europe's slow growing olive oil sector, the changes in the shares are less dramatic, from 20 percent in 1979/1 and 28 percent in 1984/86, to 36 percent in 1987.

For the DMEs to absorb the same level of net imports as at present, the growth rate of their combined production of oilseeds (in oil equivalent) should not exceed 0.6 percent p.a. between 1984/6 and 2000. In the preceding 15 years 1971–86 the production growth rate was 4.1 percent p.a. Therefore the required adjustments are very drastic. Policy measures to control production growth are already being implemented in Western Europe, e.g. under the agricultural stabilizer policies of the EC. The task is made more difficult since production of import substituting oilseeds is often considered to be an outlet for the resources made redundant by restraints in the growth of the cereals sector. Pressures in the opposite direction emanate from the high budgetary costs of these policies. Such costs are destined to increase with production at rates which will depend on the relative movements of the domestic support prices and those on the world markets for protein products and vegetable oils. Given international obligations under the GATT the EC support regimes for protein products is based on production subsidies, rather than on import restrictions, to make up for the difference between world prices and domestic target prices.

Growth of aggregate production

The preceding discussion covered the production trends and adjustment requirements of the livestock, cereals, oilcrops and sugarcrops sub-sectors. Among themselves these products cover 80 percent of gross agricultural production by value. They are used here to provide a broad indication of the agricultural growth rates which would be compatible with approximate world balance. Overall, the growth rate of agriculture in the aggregate of the countries in this study would have to slow down from 1.6 percent in the last 15 years to around 1 percent p.a. over the 15 year period from the mid-1980s to 2000, as shown in Table 4.7

Table 4.7 Growth rates of gross agricultural production adjusted for world balance

	1971–86	1983/5–2000	1984/6–2000
Western Europe	1.9		
North America	1.8		
Other Developed MEs	1.6		
Total DMEs	1.8	0.9	0.8
Eastern Europe and the USSR	1.1	1.3	1.2
Total of above country groups	1.6	1.0	0.9

Maintenance or slight acceleration of agricultural growth would seem to be in order in the CPE region as a whole. These projections appear modest in the light of ambitious targets for production growth to be found in plans and policy statements of some CPE countries. Thus, for the USSR Aganbeguian (1989) states that 'the implementation of the new agrarian policy adopted at the March plenary session of the CPSU Central Committee will make it possible to increase the production of food at the annual rate of 5 percent'. The discussion in this and the preceding chapters provide the context within which the prospects can be assessed. This involves issues of both feasibility and need for such growth rates over the medium to long term. As noted, more efficient use of feed resources would attenuate the need for rapid growth of gross output. On the other hand, import substitution would provide scope for production growth to be in excess of that of domestic demand in the medium term. At the other extreme, Czechoslovakian assessments, implying a growth rate of gross agricultural output of 0.9 percent for 1985–2000, are in broad agreement with this study and probable longer-term needs.

Concerning the Western countries as a whole, a slowdown in production growth rates would be required. How the total burden of adjustment may be apportioned between the major DME country groups is and will continue to be a key policy issue of international agriculture. The last few years have witnessed efforts to slow down the growth of production; cautious and gradual ones in Western Europe, e.g. in the dairy sector of the EC; bolder and abrupt ones in the USA, e.g. cereals in both 1983 and 1987. There is certainly great need for policy coordination aimed at minimizing the social and economic costs of adjustment, with due recognition of the legitimate development and trade requirements of other countries, both developed and developing ones. The key policy issues related to these aspects are discussed in the remainder of this book.

PART II

MAJOR POLICY ISSUES
AND OPTIONS

5 Controlling agricultural supply

Introduction

The projections outlined in the previous two chapters indicate that for most countries of Western Europe and North America agricultural policy will continue to operate in an environment of significant overcapacity in the agricultural sector. Even with the changes in agricultural policy in recent years, the potential supply capacity of the sector under a continuation of these policy regimes will exceed likely demand. These countries are therefore faced with the need to reduce further the existing policy incentives to increase agricultural production while at the same time limiting the adverse effects on other agricultural policy objectives through the innovative use of other, more selective, policy instruments. This chapter examines the options open to the main West European and North American countries which are required to redress market imbalances by reducing the growth rate of agricultural production. The issues relevant to countries which still seek to increase the growth rate of agricultural production are analysed in Chapter 6. Later chapters examine policies which can be used to pursue the other objectives of governments in the agricultural policy area.

Reduction in producer prices

Government support for domestic agricultural producer prices, often introduced in pursuit of farm income and other social objectives, has contributed to the increase in agricultural production capacities. Most governments in Western Europe and North America have recognized that lower price support to farmers is an important element in restoring balance and that there is no long-term future for an agriculture largely divorced from market trends. The stabilizer proposals adopted in early 1988 by the European Council are a case in point. The object of lowering prices is primarily to reduce production by encouraging resources to move out of agriculture, but it also contributes to the reduction of imbalances in a number of other ways. Demand is stimulated to the extent that final consumption responds to lower prices, new uses for available output become more attractive, and markets previously lost to import substitutes or non-agricultural commodities can be regained. In exporting countries where farm prices are maintained above world levels, budget costs are directly reduced. Lower prices also reduce the cost of absorbing redundant resources into new uses, including recreation, amenity and afforestation.

Agricultural prices have been reduced in a variety of ways. In many countries support levels have been held constant in nominal terms or increased at a rate less than inflation, so that guaranteed prices have been reduced in real terms. In other countries regulations for support eligibility have been tightened, e.g. by raising the quality standards for support in the EC cereals and beef regimes. In the USA participation in the Acreage Reduction Program is obligatory for farmers who wish to benefit from price guarantees. Some countries have introduced a more automatic link between producer price determination and economic variables indicating the extent of market disequilibrium. This is now a key feature of the stabilizer mechanisms in many EC commodity regimes. For commodities such as cereals, oilseeds, sheepmeat and wine, guarantee thresholds have been set and if production exceeds these thresholds support prices for that or the following year are reduced. In the USA the 1985 Food Security Act gave the Secretary of Agriculture discretion to lower loan rates faster than anticipated in the legislation in particular market circumstances. Another approach has been to reduce prices for delivered output above a certain quantity, the so-called quantum approach. Yet another alternative has been the imposition of levies on market supplies to contribute towards the cost of surplus disposal, e.g. milk, sugar, cereals in the EC. As a result of these measures singly or in combination, prices received by farmers have been falling significantly in real terms in recent years, e.g. they declined by 15 percent between 1980 and 1986 in the EC-10.

The major shortcomings of a price reduction policy are the uncertain and slow response of farmers (implying that the initial reduction must be severe to have any effect in reducing output growth), and the adverse effects on the level of farm incomes and on the value of assets owned by farmers. Large, efficient farms, which often account for a large part of aggregate production, may remain competitive, even following significant declines in prices but small marginal farms may find it difficult to survive, although this may not have an appreciable effect on total output (see also Chapter 12 on the competitiveness of European cereals production and estimates of production declines if prices were lowered to world market levels). Lower prices may also contribute to the withdrawal of land from farming, particularly in less favoured marginal areas, with detrimental consequences in regional development terms and for rural population. Moreover, a distinction should be made between the effectiveness of price falls, or indeed of other policy instruments, in reducing production of any given commodity and effectiveness in reducing total agricultural output. The latter is much more difficult to achieve than the former. Policy formulation must take into account commodity interdependence (substitutability and/or complementarity in production and/or consumption) in order to avoid achieving results in one commodity sector at the expense of worsening imbalances in others, e.g. milk and beef, cereals and oilseeds. There are numerous examples of policy interventions in the different commodity sectors which work at cross purposes with each other. For example, the US 1985 Food Security Act aims to reduce production of milk (by paying farmers to cease production) but at the same time its other provisions encourage milk production (by reducing the price of feed grains while continuing high milk support prices) and discourage demand (by raising the milk/soybean price ratio and thus encouraging substitution of margarine for butter in consumption). A recent EC Study (CEC 1988d) finds that sizeable benefits can be obtained from policy reform aimed at lessening existing inconsistencies (or disharmonies) among commodities in the agricultural support policies in the EC and the USA.

The price decline necessary to achieve a given reduction in production is determined by the price elasticity of aggregate agricultural supply. As an example, if the medium-term supply elasticity is assumed to take a value of 0.3 (the value used in the OECD Trade Mandate study, OECD 1987), to reduce production by 3 percent requires a 10 percent cut in prices. This assumes that all other conditions remain unchanged, e.g. that no further technical progress would take place (see below). There is considerable controversy on the magnitude of this elasticity, although it is generally accepted that the figure will be higher, the longer the time period the sector has to adjust (for more discussion see FAO 1987). There are also reasons why the response to falling prices may be less flexible than the response to a price increase of a similar magnitude. Many farmers are virtually trapped at certain production levels because of large fixed and small variable costs. The more specialized the farms in particular products, e.g. tree crops, the less the flexibility to respond to price falls by reducing production. To sustain their income levels, continued production is the most profitable alternative. Lower prices may encourage farmers to better exploit the production potential of existing and new technologies, lowering the use of inputs (and thus costs) per unit of output. Also, falling land prices feeding through to lower rents, as well as lower prices for feed and animals, will buffer the profitability consequences of falling commodity prices, at least for some farmers.

Further, unless the underlying growth in productivity is checked the restrictive effect of lower prices may well be masked. For example, despite the fact that real agricultural prices in the EC-10 fell by 11 per cent between 1975 and 1980 and by a further 15 per cent between 1980 and 1986, agricultural production increased by 16 per cent in the first period and by 10 per cent in the second (CEC 1988a). These historical associations between the falling real producer prices and increasing production are sometimes erroneously ascribed to the existence of a 'perverse' supply response in agriculture, i.e. lower prices encourage farmers to produce more in order to maintain incomes. The true situation is that production growth would have been even higher but for the falling real prices (for a thorough discussion see Harvey and Whitby 1988: 166–71). In the longer run, there is more scope for reducing resources employed in agriculture and thus for price falls to be more effective. This is so because the lower profitability of agriculture would reduce the rate of capital investment (and thus the rate at which innovations embodied in new capital equipment are adopted by the sector) as well as the resources (both public and private) devoted to agricultural research. Related to the above is the question whether, as some believe, technical progress in agriculture is largely exogenous, fuelled by autonomous advances in basic scientific understanding, and thus not closely related to profitability. If this were so, reducing profitability would not greatly affect the rate of discovery of more efficient techniques. However, their commercialization and rate of adoption of such would still be influenced by economic conditions such as remunerative rates of return. All these processes however require time to take effect, and in the short run productivity growth is difficult to influence by price policy measures.

The extent to which lower prices affect per capita farm incomes depends on how fast prices fall in comparison to the growth of labour productivity and on how land, labour and capital markets operate. Because of the above-noted increases in production and productivity even when real prices decline, aggregate farm income need not decline and per caput farm income may increase due to labour outflow from agriculture. In the EC-10 net value added per person in agriculture increased by 5 percent in real terms in 1980–6 (CEC 1989a: 53) when, as noted above, real producer prices fell by 15 percent. However, it is impossible to avoid some adverse effects. Even if the labour element in farm incomes is essentially set by the opportunity cost of farmers' labour and management skills in the non-farm sector, and this element is unaffected by the presence or absence of support, the income of farmers who own their land (and these are the majority in the market economies of this study) contains a substantial rent element which would be adversely affected by a fall in farm prices. Further, in the more usual situation where a considerable gap can exist between the labour element of farm incomes and alternative wage opportunities, due to resource immobility, this element too can be depressed by price reductions.

Marketing quotas

In the light of the relative unresponsiveness of agricultural production in the short-term to lower prices, and also because of the potential adverse effects of lower prices on farm income, governments have in general been unwilling to place sole reliance on this policy instrument to redress market imbalances. Hence, other measures to restrict production growth have increasingly been adopted and some governments even place the major emphasis on supply management measures. Marketing quotas for individual commodities in surplus are one such measure. A marketing quota directly limits the quantity of a commodity which may be sold. Different mechanisms have been used to achieve this result. For example, some countries allocate specific sales quotas to individual farms. In some cases more elaborate 'quantum' schemes allow for decreasing levels of price with successive tranches of output (e.g. EC sugar regime). In the case of the EC milk quota regime, sales above a farm's or creamery's delivery quota are subject to a tax ('superlevy') which effectively makes above-quota sales uneconomic. All quota systems face the problem of what basis to use to make the initial allocation of quotas, and how subsequently to monitor the output from individual farms. Monitoring is easy where all the product is sold off the farm via a central agency, e.g. for milk and sugar beet, but it is almost impossible for farm-gate sales and for products (e.g. grains) which are an input into other farm enterprises. Marketing quotas are thus at best relevant to a subset of agricultural production, and if applied to these products are likely to shift the problem of surplus production to 'uncontrolled' products as farmers diversify production.

Provided the quota policy is sufficiently restrictive, it has the advantage of being a certain method of controlling sales of a commodity. It permits some freedom to discriminate among farmers, allowing certain producers to be preferred while pushing the burden of adjustment on to less favoured producers. For example, Switzerland fixed an upper limit of milk per hectare which can be delivered in its milk quota scheme, thus penalizing more intensive producers. While such discrimination may permit the achievement of social or environmental goals, it has costs in terms of misused resources and distortions in the development of the industry. Some governments have softened the impact of a quota system on farm incomes by raising producer prices, given that the production response to higher prices is now prevented. In Switzerland the milk price was increased by about 25 percent between 1976 and 1983 following the introduction of a quota system in the previous year. If the quota system is intended as a transitory mechanism to assist the balancing of supply with demand, raising prices undermines their adjustment role by making it more difficult to remove them ultimately. Moreover, higher prices ultimately transferred to the consumer may aggravate the problem of surplus production even under a quota regime if they discourage consumption, e.g. through substitution in consumption of other products whose prices are not raised at the same time (margarine for butter being a prime example). Another way of limiting the adverse income effects of restricting production through a quota scheme is for the government to buy out the producer's right to produce at the guaranteed price. Examples include the US and EC's dairy herd disposal programmes which compensate farmers for giving up milk production on their farms for a specified period. Buying-out programmes often have high budget costs in relation to the production reduction achieved, and may cause a temporary spill-over of problems into other markets (e.g. surpluses in the meat market from a dairy herd reduction scheme).

A quota system tends to 'freeze' existing production patterns and structures, and may raise concern about the consequences for efficiency of an increasingly rigid agriculture. Normal structural change in response to a change in relative costs may be obstructed, and producers may be prevented from fully exploiting economies of size. It is possible to design quota programmes to avoid excessive costs in this regard. Some economists advocate making quotas marketable in order to retain flexibility, although any efficiency gains from this approach are bought at the expense of reduced support to farm incomes. This is because those buying quotas effectively pass the capitalised value of the income support they receive to the initial quota holders. The EC's dairy sector is a case in point where structural rationalization has continued under a quota regime, partly because it permits the transfer of quota when it is accompanied by the sale or lease of land, and partly through the administrative re-allocation of quotas made available by exiting farmers. More generally, quotas tend to increase the capital requirements for new entrants into farming. The net macro-economic effect may be to reduce the savings rate of the economy if the sellers of the quotas do not save their windfall gains.

The appropriate use of quotas, within the wider context of quantitative limits to production, is at the heart of the debate over supply control policies in agriculture in the western market economies. It assumes particular importance in the quest for policies which would enable governments to continue to pursue in varying degrees some cherished domestic policy objectives while working towards the establishment of a more market-oriented agricultural trading system in the context of the Multilateral Trade Negotiations on agriculture (see Ch. 12). There is perhaps a paradox that, at the time when many governments are committed to reducing the scope of bureaucratic controls in economic activity through policies of de-regulation, the use of quantitative controls in agriculture has tended to increase. The explanation for this paradox, as pointed out above, lies in the apparent unresponsiveness of agricultural production to moderate price cuts, given the underlying trends of productivity increases, and in the greater potential offered by quotas and related measures to target both production control and other objectives (for options within the quota framework see Harvey 1988). A role for quota policies as a supplementary device to control production has already been widely accepted. There is less agreement over whether quota policies should be regarded as a permanent feature of agricultural policy in market economies and given the dominant role in supply control, thus allowing price policy (as in Switzerland) to focus solely on income support. High-income countries which are basically self-sufficient in temperate-zone agricultural production and where agriculture is important mainly for rural population and regional development reasons (such as the Nordic and Alpine countries) tend to be the main supporters of a major rather than supplementary role for supply management.

Input control policies

The basic cause of production exceeding remunerative outlets is that there are too many resources engaged in agricultural production. One approach to the over-production problem is therefore to try to limit directly resource use in farming. In all countries, changes in resource use in agriculture have followed broadly similar trends, albeit with some important differences between countries. Generally, labour input has fallen rapidly and there has been a slight reduction in land use, while the use of capital and intermediate inputs has steadily increased. Table 5.1 shows patterns of input use for the EC-10.

If the same input–output relationships hold, slower production growth in the future would translate into faster declines in land and labour employed in agriculture and lower growth in the intermediate and capital inputs. How might these resource adjustments be brought about by acting directly on input markets? For countries which have input subsidy policies (including access to cheaper credit) in place, the first course of action is to abolish these subsidies. Thus some governments have restricted or eliminated investment aids to farmers, while others have removed subsidies on variable inputs or have reduced their investment in the public agricultural research and extension system. Additional restrictive measures might include attempts to redirect land into other uses or into retirement, buy-out schemes such as cow slaughter programmes, the uprooting of fruit trees and farmer retirement

Table 5.1 Input trends in agriculture in the EC-10, 1965–85
(annual growth rate in percent)

	Labour	Capital	Land	Intermediate inputs	Total
Belgium/ Luxembourg	−4.1	2.1	−0.8	1.8	0.0
Denmark	−3.8	2.1	−0.3	1.3	−0.1
France	−2.8	3.9	−0.4	2.0	0.1
Germany FR	−4.0	2.6	−0.8	2.0	−0.0
Greece	−2.3		0.2	3.0	
Ireland	−3.7	4.4	1.1	3.1	0.9
Italy	−2.8	5.0	−0.7	2.1	−0.6
Netherlands	−1.4	3.8	−0.6	5.4	2.6
United Kingdom	−2.4	0.3	−0.1	0.1	−0.3
EC-10/EC-9	−2.9	3.2	−0.3	1.9	0.1

Source: Henrichsmeyer and Ostermeyer–Schloeder 1987.

programmes, the imposition of taxes on current inputs such as fertilizers and feeds, and more generally restrictions meeting environmental objectives, e.g. declaration of water protection zones and imposition on nitrate quotas. An attempt might also be made to limit the growth of resource productivity by restricting the use of new technologies.

A number of general considerations apply to input controls of any kind. One problem is their unpredictability. Since yields depend on a wide range of factors, including the weather, the efficacy of the limits will vary from year to year. A second complication arises because farmers use the remaining inputs more intensively, so that to achieve a given production reduction the downward adjustment in any specific input must be larger. A third handicap is that farmers will normally limit inputs by taking them from their least productive uses. For example, cutting the cereal land area by 10 percent will reduce output by considerably less than this since farmers would, unless otherwise prevented, take out of production the poor quality hectares first. Average yields will tend to rise, requiring even further input curtailment. An additional issue is the conflict between the objectives of controlling supply and enhancing rather than hindering the overall efficiency of agricultural production. Probably the most daunting problem is the administration and policing of compliance with the restrictions, and the enforcing of sanctions against those who offend.

Reduction of labour input

Labour input into agriculture is falling in both relative and absolute terms. This decline has been associated with the adoption of labour-saving technologies on farms and, to a smaller extent, with a reduction in the number of farms. Schemes to encourage a further reduction in the numbers engaged in agriculture have a role in bringing about an increase in average incomes in farming, but their importance in influencing output is more doubtful. Many of the farmers encouraged to leave are underemployed and their contribution to output is small and could probably easily be replaced by capital inputs. Often the outflow of farmers leads to farm amalgamation, in which case a greater capital base and better management on larger farms may lead to earlier and easier adoption of new technologies. If the reduction in labour input takes the form of a move from full-time to part-time farming, generally some reduction in output would be expected. However, structural adjustment schemes (e.g. pre-pension schemes) rarely allow for this possibility. Such schemes will lead to a reduction in output only if combined with a requirement that the land be diverted to a non-agricultural use or idled in some way. For example, the EC pre-pension scheme introduced in 1988 permits the payment of the retirement annuity to farmers to idle their land and divert it to some non-agricultural use such as forestry.

Land retirement schemes

Measures of this type have been developed mainly in response to increasing surpluses on grain markets. Area controls are usually in the form of the withdrawal of some parts of individual farms. The withdrawal of whole farms will generally have a greater impact on overall supply, but there are fears that whole farm programmes have adverse effects on rural communities. Compliance can be voluntary (in return for compensation as in the EC's set-aside programme in cereals) or made a condition of obtaining price support (as in the case of the US Acreage Reduction Program). The scheme, by subtracting land, will change the input proportions at the disposal of the farmers and will tend to make capital and labour resources redundant, most likely leading to their more intensive use on the remaining land, unless they can be used more profitably elsewhere. It will appeal more to farmers who can shed such resources, e.g. borrowed capital and hired labour, or who were operating with insufficient capital and labour in the first place. Land may be taken out of production temporarily, or in a more long-term fashion by diverting it to a non-agricultural use such as forestry. However, forestry is not always a suitable alternative, and the question of what to do with land taken out of production is not a negligible one. If a flat-rate incentive payment is offered, it is not necessarily the least fertile land which will be taken out of production. The outcome will depend on whether alternative uses of the withdrawn land are permitted, e.g. non-agricultural uses or growing of crops not targeted for production control. Payments under this scheme will be most attractive to land whose value in some alternative use is most nearly equal to its value in the production of the targeted crop. This land is as likely to be in one place as another, and in one productive class as another. The environmental value of land temporarily taken out of production is likely to be very marginal, while the encouragement to use the remaining land more intensively will exacerbate the environmental costs of intensive agriculture. On balance, therefore, the environmental impact of set-aside programmes, which envisage temporary withdrawal of land from production, will probably be limited or negative.

The producer may or may not be free to use the land for other products. Without restrictions, land controls for individual crops (such as the use of acreage allotments in the USA) tend to shift the impact of market disequilibria from the producer of the controlled crops to the producer of other commodities. The costs of land diversion programmes, particularly those of an annual or short-term nature, have tended to be rather high. This will be particularly the case if farmers increase their acreage in anticipation of a set-aside scheme, or in order to qualify for a larger set-aside payment when they participate in the scheme.

Taxation of current inputs

If supply control is an overriding objective, then existing subsidy schemes to farmers, e.g. credit subsidies, investment aids or subsidies on current inputs, should clearly be removed. One might go further and impose taxation on current inputs to limit their use. For example, fertilizer use has increased rapidly in most countries and has reached extremely high levels (in kg/ha) in North-west Europe and the Scandinavian countries, stimulated by favourable price relationships and by technical innovations which have enhanced the production response to its use. Some countries have introduced a tax on nitrogen, partly on environmental grounds and partly as a means of reducing production, and this option is now being more widely considered. Some Scandinavian countries have also implemented schemes to limit the use of yield-increasing feeds. In Norway, the price of concentrate feed is set (in agreement with the farm organizations) at a high level and subsidies are made available for the production of roughage in order to try to reduce concentrate use and limit yields per dairy cow. Finland has imposed an excise duty on protein concentrates with the same objective.

The impact on production depends, *inter alia*, on the level of taxation imposed and the responsiveness of input use to the increase in price. Empirical studies suggest that a nitrogenous fertilizer tax at least equal to the price of fertilizer is necessary before a significant production response is observed. It has been estimated that in the FRG, it would take a 200 percent tax on fertilizer to reduce its use by 30 percent (OECD 1989b). The impact on production will be the more limited the more farmers have the possibility to substitute alternative sources of nitrogen (animal manures or legumes) or the possibility to substitute less nitrogen-using crops for more nitrogen-using ones (these may be precisely the changes sought on environmental grounds). The advantage of a direct tax, or other restrictions, on input use is that the adverse effect on farm incomes is less than would be the case if prices of farm output are reduced sufficiently to achieve the same reduction in land use intensity (de Haen 1989). The proceeds of the tax are available to the state to compensate farmers for this loss of income, although it would be insufficient to do this completely. In Sweden the tax has been used to help finance the cost of the export subsidies used to dispose of the country's wheat surplus on world markets. Farmer acceptance is reported to be high for this reason.

Other input control policies

One example of other input control policies are the cow slaughter schemes used, for example, in the USA, the EC and in Finland, to limit over-supply in the dairy sector. They can be targeted to particular herd sizes or to particular groups of farmers. Experience indicates that farmers are willing to participate in such schemes provided the premia offered are sufficiently attractive. The effectiveness of such schemes is determined by the degree to which milk production is reduced in relation to the number of cows slaughtered. An assessment of the 1969 cow slaughter scheme in the EC suggested that its overall effect corresponded to about half the annual milk production of all the cows slaughtered. The effects may be temporary unless other farmers are prevented from increasing their herd sizes by quota policy. A variant on direct slaughtering is to pay a withholding premium aimed either at getting producers to reduce their production or at promoting the conversion of milk producing enterprises to beef production. More generally, input quotas have the added advantage, compared with input taxes, that they are easier to target for achieving environmental objectives (see Ch. 9). They seem to be preferred by farmers and, if they are transferable, they have fewer adverse effects on farm income and equity (OECD 1989b: 39).

New technologies are being developed, in particular using biotechnology, with the potential to greatly increase yields. Already there have been ethical and health objections raised to some of these new technologies and in the United States court action has delayed testing and implementation. The banning of artificial beef growth hormones by the EC because of concern over their effects on human health is one example where proven technologies have been restricted. Although in these cases the introduction of new technologies has been opposed on health or ethical grounds, such action might also be proposed for purposes of supply control alone (for discussion of issues related to the impact of biotechnology see Ch. 7). However, to restrict the use and adoption of cost-reducing technologies other than on ethical, environmental or health grounds where clear additional social costs can be identified would be a marked departure from traditional attitudes to the value and use of new technologies in Western societies. It assumes that governments are unable to design adjustment policies to redistribute the costs and benefits of technical progress in ways which maintain the potential increase in living standards which new technologies provide. On a more practical level, unilateral action would have only a limited effect if trading partners did not comply, and, given the pressures for growth in other regions, the ability to reach such an agreement must be doubted.

Action to limit farm size and structural change in agriculture can be an indirect way of slowing down the pace of technical change. A maximum limit on herd size (e.g. legislation introduced in Finland in 1979 limited the maximum size of dairy herds to 30 cows) or on the permissible number of livestock units per hectare restricts the development of intensive livestock husbandry. In the Netherlands legislation setting down land requirements for the disposal of animal manure is an effective restraint on the expansion of intensive livestock (pig) production there. To the extent that such measures go beyond what is necessary to limit the negative environmental effects, the social cost of producing agricultural output will be higher than it otherwise would be. On the other hand, the implied inefficiency is not necessarily greater than that inherent in alternative methods of supply control.

Conclusions on supply limitation measures

A key feature of the agricultural policy environment for most countries of Western Europe and North America in the next decade will be the need to reduce the policy incentives to increase agricultural production given the limited prospects for demand growth. Two characteristics of this adjustment process make it a particularly contentious one. One is that the political costs of policy reform, already high because of the adverse distributional consequences for producer groups, are further increased if the process is not seen to involve mutual and balanced adjustments by all countries which have agricultural support policies in place. This is the issue at the heart of the Uruguay Round of trade negotiations involving agriculture, and is considered further in Chapter 12. The second is that different mechanisms to limit supply growth have different implications for the way the costs and benefits of lower agricultural support are distributed among the major domestic interest groups and between domestic groups and third countries. Some of these issues have been considered in this chapter.

In general, reducing policy-supported farm prices puts most of the burden of adjustment on domestic farm groups while leading to potential gains for domestic consumers and taxpayers. Increased real incomes and employment may follow for society as a whole, although the potential losses in social welfare if lower prices lead to more rapid abandonment of rural areas also need to be taken into account. To avoid as far as possible these negative distributional and social consequences, reduced support prices should be accompanied by the strengthening of compensatory policies in these areas.

Some small, high-income countries where agriculture is maintained largely for social reasons will continue to prefer supply management measures rather than price reductions. The major objection to sole reliance on lower support prices to redress market imbalances lies in the apparent unresponsiveness of agricultural production in the short run to lower prices. Governments in the past have therefore combined prudent price policies with quantitative restrictions on output and by policies which expressly seek to reduce the use of particular production factors. For any given degree of production restraint such policies alleviate the downward pressure on farm incomes and asset values at the expense of somewhat higher social costs and lower incomes in non-farm sectors of the economy. As indicated earlier, one of the key questions in the agricultural policy debate in the market economy countries is whether such 'indirect' approaches to supply control should be seen as temporary and supplementary or as a permanent feature of agricultural policy.

The successful implementation of supply limitation measures may well depend on institutional changes in the policy formulation process in many countries. For example, although the EC as a whole is very conscious of the need to lower incentives for increased production, individual EC member countries have an incentive to expand production because, viewed narrowly from the standpoint of agricultural policy costs alone, the financing rules of the Common Agricultural Policy allow them to 'externalize' the costs associated with increased output. In some countries, it may be appropriate to increase the role of finance and other ministers in agricultural policy, as has been done recently in the EC, in order to bring wider social interests to bear in agricultural decision-making. In all countries, greater transparency in decision-making would contribute to more efficient and cost-effective decisions.

6 Promoting agricultural growth and productivity

Production objectives

While for many countries in North-west Europe and North America overcapacity in the agricultural sector is the principal production problem which must be addressed in the immediate future, for other countries, particularly the Centrally Planned Economies (CPEs) and some Mediterranean countries, agricultural policies will continue to aim at promoting agricultural growth. In some countries this will reflect the fact that the weight of agriculture in the economy remains relatively high, and balance-of-payments problems influence policies of import substitution and export expansion. In other countries, domestic demand growth will continue to be a dynamic element requiring significant increases in supplies, e.g. in some Mediterranean countries.

Production expansion is least likely to conflict with other policy objectives where it takes place on the basis of increased productivity and lower levels of price support. Increased agricultural production efficiency is sought by all countries both as a means of achieving sectoral objectives such as improved farm incomes, lower consumer prices and increased market share, as well as for its contribution to macro-economic objectives such as faster agricultural and economic growth. Although this makes sense for each country individually, the effect of all countries pursuing this goal simultaneously is to make the transitional problems of agricultural adjustment even more difficult.

Traditionally, different policy instruments have been used to pursue growth and efficiency objectives in Western and Eastern economies. In market economy countries, price policy or equivalent policies (e.g. production-linked deficiency payments) have been the most important instrument used to stimulate and guide production, while public investment and input allocations, within the framework of centrally set and compulsory plan targets for production units, have played a more important role in the CPEs. These differences are narrowing as CPEs make greater use of market-based incentive systems in their agricultural planning. In addition, both groups of countries have made extensive use of investment aids, credit policy, taxes and subsidies to promote agricultural investment. The critical role of agricultural research, education and extension in both regions is considered separately in Chapter 7.

Pricing policies

Pricing policies have been playing an increasing role in the countries of

Eastern Europe and in the USSR, especially since the early 1980s when several major increases in producer prices and complete overhauls of pricing systems were instituted. These changes in producer prices and pricing systems reflect a drive to greater market-oricntation and to greater accountability. They are part of the wider process towards policy reforms aimed at greater decentralization in decision-making. Despite this general trend in the CPEs towards greater reliance on pricing instruments, there are significant differences in their role and importance between countries. In particular, an important distinction must be made between those countries (such as the GDR, USSR, and Czechoslovakia) where the determination of producer prices was up to quite recently largely divorced from consumer prices and those countries (such as Poland and Hungary) which are moving rapidly towards prices freely determined by demand and supply. Thus, in certain sense, these latter countries are allowing market forces to determine agricultural prices to a degree which is well beyond that prevailing in many market economies of Europe. Additionally, price policies, particularly in large countries like the USSR, have to cope with the problem of widely differing production and transport costs in their different regions because of their wide diversity of agro-ecological conditions and of distance from the main markets. In the current reforms in the USSR, the method of accounting for such diversity is being changed. The reforms reduce zonal price differentation and, in market-like fashion, land quality differentials will be reflected in differentials in the land rentals charged to farm enterprises in the form of tax rather than in the form of higher or lower producer prices.

A key characteristic of agricultural pricing policies in the CPEs in the past has been its dual character. During the 1960s and 1970s producer and consumer prices were progressively disconnected as each served a different policy objective. Producer prices were centrally determined, in most cases independently of international market prices. They bore no necessary relationship to real costs and were often insufficient to provide an adequate income for producing units. This was especially the case during the period of compulsory deliveries at low fixed prices serving the objective of resource transfer out of the agrarian sector. Consumer price policies, on the other hand, were oriented to the objective of low and stable retail prices. Consistency between the pricing decisions at various parts of the food production and marketing chain was pursued through transfers and subsidies from the state budget. With the substantial increase in producer prices in many CPEs in the early 1980s, the share of these subsidies in state budgets increased dramatically (retail price subsidies increased to about 14 percent of the state budget of the USSR in 1986 and about 13 percent in the GDR).

One objective of the reorientation of pricing policies in CPEs is to rationalize subsidies. As the desire to eliminate subsidies has been accompanied by a strengthening of producer incentives through higher prices at the farmgate level, it follows that the objective of subsidy elimination can only be achieved if retail price levels too are allowed to rise. Several CPEs, e.g. Bulgaria, Hungary, Poland and Romania, have followed this course in the 1980s. Given the high share of food (including beverages) in household expenditure, the impact on real household incomes should not be underestimated and this factor may constrain the speed with which the objective of subsidy elimination can be pursued (see also discussion in Ch. 3 and 10).

In the longer run, the reorientation of pricing policies has a more fundamental role to play in bringing about a more rational use of production factors in CPE agriculture and the economy in general. By reflecting the true social costs of scarce resources to enterprise managers, a more market-oriented price policy will help to direct resources to their most efficient uses. To achieve this end, however, price policy requires a number of pre-conditions. In those countries which still maintain central planning as the main approach to managing the economy, the effectiveness of price policy is linked to the willingness to decentralize and provide greater scope for local managerial autonomy. Production factors themselves, including labour and capital, must be priced to reflect their scarcity values. Another issue is the extent to which policy-makers can look to the world market for guidance in price formation. A recent policy statement from the USSR states that 'preparations are being made for the comprehensive price reform aimed, *inter alia*, at achieving comparability and closer levels of domestic and world commodity prices' (UNCTAD 1989). Using international market prices as an 'objective' anchor becomes particularly difficult when the exchange rates prevailing in some countries are far from reflecting the scarcity value of foreign exchange. Of relevance is also the question whether distorted international prices, because of export subsidies and agricultural trade protectionism, are a good or bad guide. The key criterion here is not so much that prices are distorted but that there is increased uncertainty as to what future world market prices will be, both levels and degree of instability, because agricultural support and protection policies in major trading countries may change, including as a result of the Uruguay Round of the GATT Multilateral Trade Negotiations. Additionally, the use of world market prices as a guide for setting domestic prices will require corresponding reform of foreign trade policies, so that changes in demand and production following price reform can be reflected in changes in imports and exports. This is among the reasons why agricultural price reforms must be seen in the wider context of overall economic reform involving, among other things, issues of currency convertibility and enhanced access to foreign markets by producers and consumers.

Thus for the CPE countries the creation of an internally coherent system of incentives and economic instruments will be a crucial issue. It is not enough simply to make particular instruments more active; even more important is that they provide a consistent set of signals and incentives to producers. Pricing policies are, however, only a part of a package of incentives relating to all aspects of agricultural production. The effectiveness of producer price incentives will depend crucially on the extent to which producers can react to them. This will require, among other things, relaxation of those policies which, explicitly or implicitly, impose centrally-set levels for both prices and quantities for parts of production. An important aspect of the introduction of a price-based incentive system is that it reinforces a wide range of other market-oriented policy instruments, such as the opening of new market channels (e.g. direct sales to consumer cooperatives instead of state procurement bodies) and the granting of preferential access to inputs (mechanization, chemicals, etc.) in the case of good performance.

In the USSR a scheme was recently introduced (August 1989), whereby state and collective farms can be paid in hard currency for above-average deliveries to the state of wheat, pulses and oilseeds, provided production of *all* grains, pulses and oilseeds exceeds average production of earlier years. The hard currency proceeds can be used to import production requisites or consumer goods, in effect giving recipient farmers preferential access to such imports. The evidence is that such preferential access has a high price. If effectively implemented, the scheme could, therefore, provide a hefty increase to the purchasing power of farm sales proceeds from wheat, pulses and oilseeds, as well as to real producer prices in practice. The cost of this implicit subsidy to grain producers would be borne by the rest of the society, including the producers of other agricultural products, whose access to imports would be correspondingly curtailed, unless foreign exchange availabilities increased. The scheme is indeed meant to save foreign exchange on food imports by promoting import substitution, with the state buying wheat and oilseeds from domestic producers rather than importing them. Import substitution would occur, however, only if production of these commodities increases and that of other products does not decline, from what would have been otherwise, as a result of the relative price shift in favour of wheat, pulses and oilseeds. Otherwise, the scheme, by changing relative prices in the different markets, could simply divert to state procurement supplies from other uses, e.g. feed, with the consequent decline in livestock production under unchanged feed conversion rates and livestock prices. This could indeed happen, and import substitution would be achieved through demand curtailment, unless the state were to re-supply at non-increased prices the markets from which supplies were originally diverted.

In market economy countries price support policies have contributed to agricultural expansion in the past by encouraging the use of more resources in agricultural production, greater adoption of technological advances and the promotion of more research than would otherwise be the case. If price policy is used as an active element in agricultural growth policy, this implies a willingness to extend and increase levels of protection to the agricultural sector. In view of the stated commitment of governments in these countries to reduce agricultural protection, and the greater awareness of the costs of price support programmes, this route to raising farm output is likely to be used less by market economy countries in the future, and instead these governments may well concentrate on other policies designed to raise the productivity and competitivess of their farming sectors.

Public investment policies

Public investment allocations have been one of the most important instruments to promote agricultural growth in the CPEs, particularly in the

USSR, even to the extent that one might speak of over-reliance on this single instrument of policy. Table 6.1 shows the shares of total investment going to agriculture. The USSR is reported as devoting 26 percent of total material sphere investment to agriculture in 1981–5 when the sector contributed only 10 to 12 percent of total Net Material Product (NMP). This cannot be translated into an agricultural investment rate (agricultural investment percent of agricultural NMP), due to unavailability of data on the total amount of investment in all material sphere sectors. Other data indicate that gross fixed productive investment in agriculture in the USSR was 43.3 percent of agricultural NMP in 1980, which is very large and indicates over-capitalization of the sector.[1]

Data on capital productivity (net output per unit of fixed assets) indicate that in agriculture it is half as large as in the economy as a whole. To judge from the relative shares of agriculture in NMP and in total investment, similar conditions prevail also in the GDR, Czechoslovakia and Poland. In general, the implied high investment rates in agriculture underline the excessive reliance of some CPEs on this single policy instrument; more investment not supplemented by other appropriate policies does not necessarily make for more efficiency. In the USSR there is increasing recognition that simply putting more resources into agriculture will not by itself solve the food problem.

Table 6.1 CPEs, shares of agriculture and capital productivity

| | Shares of agriculture in material sphere total, % | | | | Net agricultural out put per unit of fixed assets | |
| | Investment | | NMP | | (total economy =1.0) | |
	1976–80	1981–5	1976–80	1981–5	1971–75	1981–85
Bulgaria	13.9	11.0	21.5	14.6	1.8	1.1
Czechoslovakia	17.7	20.0	7.9	7.9	0.7	0.5
German DR	13.4	11.6	8.8	8.0	0.7	0.6
Hungary	18.9	19.3	18.7	19.5	1.3	1.1
Poland	25.1	28.2	15.3	17.6	0.6	0.6
Romania	17.0	20.4	20.3	15.7	1.6	1.1
USSR	27.5	25.8	12.5 *	10.2 *	0.9	0.5

Source: UN/ECE 1987, Tables 3.7.3, 3.7.6.
* UN/ECE (1988: 225, 227) reports 12 percent for 1981–5 at 1976 prices and (for 1985) about 20 percent at current prices.

[1] Since the implied investment rates are so high, serious doubts may be raised concerning their comparability with the investment rates of western countries. One reason advanced by the ECE concerns the prices used to compute the values of agricultural output and current inputs (hence of agricultural NMP) and those of capital goods: 'Output prices for agricultural products have been maintained at low levels relative to other sectors in all seven countries under review — and despite price revisions they remain so in most to this day. Agricultural input prices, on the contrary, have tended in the past to be relatively high compared to other sectors' (UN/ECE 1987: 187).

With the recent reorientation of agricultural policies in the CPE, countries investment policy is also being changed in order to make investments more efficient and to strike the right balance between investments in agriculture and investments in the rest of the agro-industrial complex. There is a growing emphasis on investment in the rehabilitation of fixed assets and a tendency to give agro-industrial enterprises at the local level more say in the assessment of investment projects and the distribution of resources. It is interesting that in the context of farm policy reform in the USSR the possibility of reduction in investments in agriculture is at least considered as a policy instrument towards raising productivity, in parallel with higher emphasis on investment in transport, processing, storage and rural development.

More effective management of the agricultural industry will be of crucial importance in bringing about further improvements. Within the large-scale socialized sector, efforts will be made to integrate better, both at the production level and at the administrative level, the various branches of the agro-food complex, e.g. creation of Ministries of Agriculture and Food which replace a whole series of specialized branch ministries. Some of these measures did not meet with success and were subsequently reversed or replaced by new organizational arrangements. It is, thus, unlikely that administrative reforms alone will bring about the improvements in investment productivity which are sought, though they can contribute to increasing productivity. A case in point is the need for such reforms to reduce the degree of uncertainty surrounding the permanence and stability of arrangements defining the rights and obligations of producers under the different systems, e.g. under contract farming or leases of land and other farm assets. This is particularly important since increased productivity will depend on improved management at enterprise and sub-unit level. This is in turn intimately related to the certainty as to the commitment to a more market-oriented policy with prices reflecting real costs and to the development of enterprise autonomy permitting the rewards, and sanctions, of successful or indifferent enterprise performance to apply.

Apart from high overall rates of investment, an early and characteristic feature of agricultural investment allocation in CPEs was the distinct preference for the reorganized large-scale state farms and cooperative units. Directed credit allotment and controlled allocation of scarce production inputs expressed this policy preference. Among the recent policy initiatives considered in the USSR are provisions that would allow individuals and groups to lease land, equipment and buildings for up to 50 years and, under certain conditions, to transfer the lease through inheritance. Other provisions would remove discrimination in the supply of inputs among leaseholders and peasant farmers on the one hand and state and collective farms on the other. Similar non-discriminatory policies had already been adopted in Hungary and notably in Poland in 1981 where a reorientation of agricultural policy abolished these preferences and made available a growing supply of production inputs to the small-scale peasant farms and auxiliary household plots. The encouragement of production on household plots and auxiliary farms in several countries has been such that by the early

1980s a substantial part of the supply of labour-intensive products (e.g. certain livestock products and vegetables) came from small-scale units. Specifically, over 50 percent of total production originated in household and auxiliary farms in 1980 for the following commodities and countries (Research Institute of Agricultural Economics — Hungary 1982):

Potatoes — Bulgaria, Hungary, Romania, USSR

Fruit — Czechoslovakia, Hungary, German DR

Pork — Hungary

Milk — Romania

Eggs — Bulgaria, Hungary, Romania

In future, further efforts to exploit the potential of the small-scale sector will be made, for example by strengthening the mutual specialization and links between small and large farms in a particular area (for further discussion see Ch. 8).

In the USSR, private plots account for some 3 percent of arable land but they provide some 25 percent of gross agricultural production, including some 30 percent of livestock output. These shares include production from the fast expanding sector of urban garden plots. The disparity between the weights of the private plots in total arable land and in gross production, however, does not necessarily imply that land productivity in the sector of the 'private subsidiary economy' is a high multiple of that in the state and collective farms. It would be more appropriate to compare productivities on the basis of shares in value added rather than in gross output. Otherwise, the erroneous impression may be created that the policies favouring individual and group farming would result in spectacular productivity gains. Gains there may well be, though more modest ones. In practice, much of the feed (both concentrates and grazing land) used in private livestock production originates in land of the state and collective farms. When land is transferred from socialized to individual farming, the latter's crop mix may change in favour of lower-value crops now produced in the former sector, with the consequent decline in its average land productivity. This effect will be most pronounced if the transferred land substitutes mainly for feed now provided by the socialized sector. In the USSR the trend in policy reform seems to favour a symbiotic relationship between the private plots and different organizational forms of farming on land owned by the state and collective farms, e.g. contractual assignment of operations to groups of persons or families, leasing of land to such groups or individuals and official employment. Recent policy initiatives encourage such symbiotic relationships, for example by abolishing centrally set limits on the size of private plots and allowing state and collective farms to count towards their plan fulfilment goals the livestock raised under contract with the private plots.

In the market economies public investment plays a more modest but still important role, particularly in areas where there are significant externalities or public good characteristics such as plant and animal health, water and irrigation investments, land drainage and improvement, and research and extension. In many countries natural conditions need improvements on a scale which requires the state to bear the expense of their execution: irrigation, erosion control, reafforestation, conversion and improvement of land. Irrigation plays a particularly important role in many Southern European countries and in the United States (where usage of surface water is rapidly approaching its maximum potential). Benefits of large-scale works financed by public funds are often treated as (nearly) free goods. From both an equity and resource allocation standpoint a greater contribution, particularly to cover costs of administration and upkeep, should be required in the future from the beneficiaries.

Another area where public investment has been important in both CPEs and market economies is in improving livestock production, in the form of breed improvement programmes, the encouragement of artificial insemination, the provision of veterinary and disease control services as well as research and extension, the development of fodder production, and the improvement of marketing and slaughtering facilities. The emphasis in these programmes in the future must be on increasing their cost-effectiveness through improved management. The possibilities of increasing the contribution of farmers to these services through sales levies or charges should also be investigated. Building up the capacity for sustained agricultural innovation through investment in public agricultural research and extension programmes, as discussed in Chapter 7, has a key role to play. Both the level and management of the resources devoted to these purposes should be carefully monitored to ensure that they make the maximum impact.

Subsidy and taxation policies

With respect to the use of subsidies to promote agricultural growth, a distinction can be made between subsidies on current and on capital inputs. Examples of the former include subsidies on fertilizer, energy use and, most commonly, on irrigation water. Particularly in the late industrializing countries in the region, such subsidies are used to promote the use of modern inputs and to encourage the adoption of modern production practices. In Northern Scandinavia and the Alpine countries selective use of subsidies is made to improve the productivity of smallholders and to reduce costs for those farming in particularly unfavourable circumstances. In the CPE countries, too, current input subsidies are used to encourage the adoption of modern technical advances and to support regions with marginal, unfavourable production conditions. Additionally, input subsidies are often used in lieu of increasing producer prices. Current input subsidies can be justified on efficiency grounds if they encourage the faster adoption of profitable new innovations, but when implemented over a longer period, they can encourage an over-use of the subsidized input and distortions in the product mix.

In many countries wide use is also made of capital subsidies, such as schemes for investment aids, tax credits and subsidized credit to promote on-farm investment. Likewise, credit for agriculture is assured either by banks or specialized institutions which are in a position to grant loans on particularly favourable terms thanks to the financial support which they enjoy from the state. This policy of subsidizing credit is aimed at promoting the modernization of agriculture, both as regards infrastructure and the means of production. Such schemes can be justified by deficiencies in the operation of banking and credit markets which make it difficult for farmers to obtain the finance to undertake worthwhile investments. But the dangers should be appreciated, too. Because of access to cheaper credit, farmers may be encouraged to undertake investments in buildings and land improvement which do not yield an adequate rate of return if the grant element is excluded.

In the CPEs where production factor subsidies of all kinds are widely used to promote agricultural growth, there is a clear tendency towards a less intensive and more selective use of subsidies. In the GDR, Czechoslovakia and other countries, while the intensification (increased use of modern inputs) of the agricultural production process still remains a major objective, state subsidies on purchased means of production are being abolished or reduced to encourage more rational use of these production factors. Subsidies are still maintained or being increased to promote the use of specific inputs (e.g. reduction of maintenance prices of farm machinery in Czechoslovakia and subsidies on the application of biological innovations in Hungary). In the USSR the financing of agriculture is in the process of being reformed: state-financed contributions are to be phased out and all agricultural enterprises will have to switch over to 'self-financing', based on own earnings and borrowing from banks. It is hoped that the increased costs of financing will lead to more efficient use of resources and thus to increased productivity. It must, however, remain doubtful if 'self-financing' by itself will make much difference towards higher efficiency if farms continue to operate under a centrally imposed pricing and marketing system.

In many countries the tax system is used to provide production incentives to farmers. For example, in the United States, but also in Poland, farms have been able to take advantage of special depreciation rules and investment credits for agriculture to encourage on-farm investment, and to apply income averaging to reduce the income risk in farming. Rules governing the deductability of farm losses and special tax provisions regarding capital gains give agriculture the characteristics of a tax shelter that has attracted passive investment capital from investors outside the sector. In some EC countries, agriculture is subject to a tax regime completely different from the one applied to other sectors and farm income is taxed on a notional income basis and in some extreme cases it is outright exempted from taxation. Such regimes often mean that the tax burden on agriculture is significantly lower than it otherwise would be. Thus, in 1980, the 'under-taxation' of agricultural income compared with that of other occupational categories was estimated to have amounted to the equivalent of 70 percent of public expenditure on agriculture in Belgium, 42 percent in the Federal Republic of Germany, 21 percent in Italy and 16 percent in France (CEC 1984). Recent moves towards taxation reform in some market economy countries in the region, however, tend to reduce the value of tax incentives given to agriculture. Taxation is also an important instrument of agricultural policy in the CPE countries. It is used, for example, to redistribute revenues from farms enjoying more favourable natural conditions of production to those located in more marginal areas. With the greater role for more market-like incentive arrangements in these countries, the scope for using taxation as an agricultural policy instrument will be enhanced.

Apart from their implications for efficient resource use, input subsidies, capital grants and tax incentives in Western Europe and North America have important structural and distributional effects. The uptake of subsidies is usually greater by larger farms. Tax incentives are most useful to those farms with a taxable income, which also tend to be the larger farms. Thus public policy to foster faster and more efficient growth may conflict with structural objectives (favouring a broadly decentralised agricultural structure) and income distribution objectives.

Conclusions

In spite of the projected increase in the degree of agricultural over-capacity for the region as a whole over the next decade, the objective of increased production will remain of importance for individual countries. Particularly for the middle income and late industrializing countries of Southern Europe, and for the CPE countries, higher and more efficient production growth will remain an important goal. In higher-income countries, the modernization of farms and the pursuit of production efficiency will continue to be encouraged as one way of coping with increased competitive pressures.

In the CPE countries the principal drive will be to increase agricultural growth and productivity by the more rational use of resources. For those countries which retain central planning as the main form of economic management, this will involve continuing the search for new management and organizational forms, the further rationalization of the subsidy system to the agro-food sector, the wider use of the price mechanism and greater reliance on enterprise autonomy and profitability. For all CPE countries, one way to more efficient resource utilization could be fuller participation in international trade. This approach is already being implemented in those countries (e.g. Poland, Hungary) which have recently introduced radical reforms involving nearly complete removal of restrictions in foreign trade transactions and movement towards currency covertibility.

In the market economies of Western and Southern Europe and North America, the case for state intervention to assist in the process of farm modernization and agricultural investment rests largely on the concept of market imperfections, i.e. instances where market forces operate either too slowly or fail to bring about a socially desirable allocation of resources to the agricultural sector. A considerable range of public measures to assist agricultural investment has been put in place on this basis, and others have been introduced on social or equity grounds. In the light of the increasing evidence that many of these measures have adverse structural and distributional effects, there is a need for a thorough examination to ensure that their continued operation is warranted.

7 Agricultural technology issues

Introduction

The key factor which over the past decades has transformed agricultural production in all study countries has been the steady flow of innovations in agricultural techniques. Technical progress has a profound impact on almost all aspects of agriculture. It provides the cost-reducing techniques which enable food supply goals and agricultural expansion to be met in an efficient way, while in the short term exacerbating both the extent of market imbalances and the transitional adjustment problems of the agricultural sector. It is the driving force behind many of the structural changes which have taken place in both Western and Eastern Europe. It is partly responsible both for causing as well as resolving many agricultural pollution concerns. It influences the underlying pattern of agricultural comparative advantage and the relative position of countries in international trade. It is viewed with increasing suspicion by consumers for changing the nature of diets in unfamiliar ways. Advances expected from emerging technologies will continue to have these impacts. Because private interests in agricultural technology do not always coincide with public interests, there is and will be a continuing need for a role for the public authorities in the agricultural innovation system.

Technological innovations have been developed through both public and private agricultural and agro-industrial research. The rapid uptake of these innovations by farmers has been encouraged by public extension and advisory programmes and by the agro-industries up- and downstream of the primary sector. All these components of the agricultural innovation system face new challenges and problems in the years ahead, and their relative roles and interrelationships need to be carefully defined by public policy.

Agricultural research

The commitment by governments in the study region to agricultural research increased substantially over the 1960s and 1970s, from a total of US$1.5 billion in 1959 to US$4.3 billion in 1980, measured in 1980 prices (Evenson 1986). Significant differences both in the levels and rates of increase in spending and manpower intensities exist between sub regions (Table 7.1). In terms of relative spending intensity, Southern Europe, Eastern Europe and the USSR appear to spend less than other sub regions, but the ratio of scientific manpower to agricultural product in Eastern Europe and the USSR is high and close to the ratio in Northern Europe (see, however, the discussion in Ch. 6 concerning possible distortions of these ratios because of the system of agricultural product valuation in the CPEs). Examining changes over time, it is clear that the big expansion in research commitments took place in the 1960s and that in a number of sub regions research commitments on a relative (though not on an absolute) basis declined during the 1970s.

In Western Europe and North America, private sector research and development expenditure in support of the food system is very significant. Expenditure by US firms alone was estimated at around US$1.6 billion in 1979, divided about equally between agricultural inputs and food marketing and distribution. The share of the private sector in total US research and development expenditure was about 65 percent in 1979, and this share has been increasing over time, as has that in the total number of agricultural scientists in public and private sector employment (Ruttan 1982, 1987).

Table 7.1 Public agricultural research commitments by sub-region

Sub-region	Research expenditures as % of value of agricultural product			Scientific man-years per 10 million (1980) dollars of agricultural product		
	1959	1970	1980	1959	1970	1980
Northern Europe	0.55	1.05	1.60	1.05	2.01	3.14
Central Europe	0.39	1.20	1.54	0.80	1.21	1.56
Southern Europe	0.24	0.61	0.74	0.93	1.17	0.96
Eastern Europe	0.50	0.81	0.78	1.44	2.97	2.84
USSR	0.43	0.73	0.70	1.38	2.37	2.34
North America	0.84	1.27	1.09	0.84	0.89	0.84

Source: Evenson (1986).

Economic evaluation of public research expenditure suggests that the rate of return on this expenditure has been very high, and that despite the growing financial commitment to agricultural research by all governments in the region there is still considerable underinvestment in agricultural research systems. In the past, expenditure may have been limited by the policy-makers' perception of excess capacity in agriculture and because of uncertainty about its future benefits. In certain countries also farmers' opposition to the price-depressing effects of research might have played a role in this respect (Paarlberg 1980). While governments in the study countries face the need to limit agricultural programme costs, the high rates of return on agricultural research expenditure suggest that restricting research expenditure is a relatively inefficient way of doing this.

Because the costs of research are increasing much faster than prices in general, the management of research systems will assume increasing importance. Some governments have increased the use of contract or competitive grant research as one way of trying to increase research efficiency, but further investigation into the factors which determine the efficiency of research systems is required. Although certain countries in Southern Europe have obtained rather remarkable results in the creation or adaptation of improved seeds (e.g. maize in Yugoslavia and wheat in Turkey), institution-building in the agricultural research area will remain an important task in this region. In the CPEs research is rather narrowly concentrated on some technological areas. It should focus more on efficient and sustainable use of resources, new management methods and output enhancing technology. In particular, research is expected to focus on biotechnology and microelectronics. An important part of the research in the CMEA countries is organized under the 'Comprehensive Programme of scientific and technological progress of the CMEA member countries up to the year 2000'.

The distributional consequences of agricultural technology advances will continue to be a focus of concern. Given the inelastic nature of the demand for food, the effect of agricultural research is to lower the real cost of food (where prices are free to fall) and thus ultimately to benefit consumers, in particular poorer people who spend a relatively larger proportion of their income on food. Under a regime of fixed minimum product prices, farmers can benefit from technological change which lowers input costs per unit of output, at least in the short run. However, large farms tend to benefit more than smaller farms. Large farms are usually early adopters of a new technology which they view as a means of increasing profits. It is more profitable for them to invest in acquiring information and also their stronger asset position enables them to take the associated risks. Lumpiness of capital inputs and economies of scale associated with new technology have been a major driving force behind the continuous increase in average farm size. Future technological innovation (e.g. emerging bio- and information technologies) may not have such a bias in favour of large farms, although their impact in slowing down the trend towards farm concentration will remain limited for some time. Indeed, most innovations that will become available and widely implemented over the next decade or so relate to more traditional technologies.

In the early 1970s there was concern that a lack of forthcoming technologies would adversely affect the growth of agricultural productivity and ultimately lead to a slowdown in agricultural growth. This view was probably too pessimistic in the first place. The dramatic advances in biotechnology have now reversed this view, although the impact on levels and costs of production, particularly through the use of recombinant DNA technologies (genetic engineering), will probably not be appreciable before the year 2000 (see below). In the United States scientists predict a quickening in the pace of yield increases in the near future, particularly for livestock products (see Ch. 4). The expected rapid pace of change has the potential for very large gains and very large losses to different groups in the food system, and could also influence significantly agricultural comparative advantage between the developed and developing countries to the possible detriment of the latter. A major critical question will be whether the release of these new technologies should be left to market forces, subject only to health and safety checks, or whether their release should be managed and controlled by public authorities in the interests of achieving wider social goals.

Because biologically oriented research was less likely to produce a proprietary or patentable product, the private sector has taken the lead in other types of agricultural research, for example related to mineral fertilizers and mechanical equipment, while public sector systems focused on biological research. With the recent legislative and technical developments permitting the patentability of biological material, private firms, particularly agro-chemical companies, are increasingly moving into areas traditionally in the public sector preserve, such as plant breeding, where they consider they can develop packages of inter-dependent technological inputs. Moreover, the growing interdependence between public and private research generates pressures for collaboration and joint ventures. The implications of growing private sector influence over research priorities and the appropriate division of labour between private and public sector research will need careful consideration in coming years.

A further question is the issue of research priorities. Because of the long lead times in research activities, the timely identification of future research problems carries a substantial premium. Emerging areas which warrant increased attention include research into lower input farming systems meant to develop (economically viable) cultivation and husbandry methods with benefical effects on the environment (e.g. lower agro-chemical input use per unit of output) and on land use (extensive methods). A related area is research on organic farming (in the interest of consumers) and free-range animal production systems (consumer's concern and concern about animal welfare). Another example is research on non-food land use (e.g. fibrous plants, agro-forestry). A further important area (in particular for Mediterranean countries and certain parts of the USSR) could be research on crop cultivation in less favourable environments.

Biotechnology

Recent advances in biotechnology raise a number of issues in the area of agricultural technologies that must be confronted by the policy-maker in the short term, although significant impacts on food and agriculture can be expected more in the longer term — beyond the year 2000. Some of these issues are similar to those posed by other technologies (e.g. the extent to which it is desirable to encourage production-raising technologies in the presence of surplus capacity in agriculture) while others contain novel elements (e.g. patent protection of man-made living organisms or approval by regulatory authorities for release into the environment of genetically modified micro-organisms).

There is no formal definition of biotechnology but for practical purposes it can be defined as 'a set of related technologies and scientific and technical methods which encompass both modern fermentation and enzyme-driven transformation processes and the new technologies arising from recombinant DNA, protein engineering and cell fusion' (OECD 1989a: 81; for a general introduction to modern biotechnology see Fowler *et al*. 1988; for examples of biotechnologies see Table 7.2). It must be underlined, however, that the novel element in most of these technologies is essentially the use of new discoveries in biochemistry and microbiology in areas of research and development (R & D) with a long established tradition, e.g. breeding or the use of yeasts and bacteria in food processing.

Table 7.2 Potential impact of biotechnology on total production* by the year 2005, EC-12

Technology	Products						
					Cattle		
	Cereals	Sugar	Oilseeds	Dairy	Beef	Pigs	Sheep
Chemical growth regulators	8.0						
Stimulatory micro-organisms	4.9	6.7	7.4				
DNA/MAB probes	5.6	6.6	7.9	6.7	7.6	6.9	4.4
Transformation of crops with genes for viral components	10.8	3.7	3.5				
Broad spectrum herbicide resistance	3.6	4.6	4.7				
Incorporation of pest-combatting genetic characteristics	5.9	6.1	5.1				
Recombinant viral vaccines				11.5	11.4	11.9	5.7
Embryo transfer, transgenic animals, endocrine interference, immunochemical control				9.0	9.9	9.0	

Source: CEC (1989d).
* Percentages by which production in 2005 would be above that which would otherwise prevail, assuming continued (but slower than in the past) trends in productivity growth from 'conventional' technology and other progress and EC policies that would ensure that prices are cut in line with such productivity growth.
Note: Impact on production obtained by combining the percentage of total farmers adopting the particular technology, the proportion of total production accounted by adopters and the impact on their yields. The increments originating in the different technologies are not necessarily additive, since the adoption of a technology may exclude or otherwise affect the adoption of another.

Of major significance for agriculture is the wider range of possibilities offered by biotechnology, compared with the conventional methods, for modifying living organisms, and doing so in quicker and more directed ways. Such possibilities include the transfer of genes between unrelated species and, more generally, tailor-made interventions affecting only the targeted gene, with the result that the hit-and-miss element of traditional breeding methods is virtually eliminated. In this way the genetic engineering techniques of biotechnology speed up significantly the development of plants and animals having desirable characteristics by reducing the number of generations required to integrate the transferred gene into a desirable genotype; unwanted material is estimated at an early stage in the breeding process which reduces the risk that other genetic traits may be affected. Modern cell tissue culture and in vitro cloning offer significant possibilities of the production and propagation of standardized species for a wide range of plants, although the former shares some of the hit-and-miss characteristics of traditional breeding.

The impacts on agriculture follow from the improved performance of the biotechnologically modified plants and animals in terms of, for example, growth rates and yields, quality of the product (e.g. oil, protein and starch content of crops, leaner meat in animals), resistance to disease and pests and tolerance to pesticides and adverse agro-climatic conditions (e.g. to frost or salt and moisture stress).

Beyond animal and plant genetics proper, biotechnology affects food and agriculture also in other ways. For example, microbial biotechnology would contribute to increase plant resistance to pests (e.g. insect-killing bacteria or, more generally, biopesticides) and improve nutrient uptake by means of genetically engineered microbial organisms able to colonize the plant roots; in the livestock sector, genetically engineered hormones like bovine somatotropin (BST) can improve feed efficiency in milk production, and embryo transfer technology (using low-grade cows as foster mothers for embryos from high-grade animals) can contribute to the growth of high-grade livestock, at the cost, however, of decreased genetic diversity. In animal health, techniques like the use of monoclonal antibodies lead to improved diagnostics and disease treatment while biotechnology holds promise for new and improved vaccines for controlling animal diseases. In the feed sector single cell protein (a bacterial protein derived from hydrocarbons or biomass raw materials) is an additional source of protein for animal feed, although it is not presently competitive with other sources; biotechnologically produced feed additives can upgrade feed quality and digestibility, while the new techniques may, in the longer term, provide means of modifying rumen fermentation or altering rumen micro-organisms. In the food processing sector, biotechnology applications (fermentation technology, food microbiology, enzymology) underpinned the spectacular growth of the high fructose corn syrup (or isoglucose) industry in the USA (see Ch. 3) and hold promise for increased efficiency of the industry and the production of foods with improved quality (taste, flavour, freshness) and safety.

In assessing the impact of existing and new advances in biotechnology on the issues confronting European agriculture, one aspect seems to be of predominant importance: the countries which are sufficiently advanced technologically and industrially to pursue biotechnology R & D and its applications are precisely those which have problems of insufficient demand growth for their farm output coming from already applied and further progress and diffusion of conventional technology and/or are reluctant to see the rate of decline of agricultural employment accelerate further and budgetary costs to support higher output increase further. It is estimated, for example, that BST adoption in the USA could increase costs of payments to farmers by $90 million in 1996 if current support levels were maintained (Molnar and Kinnucan 1989: 10). Table 7.2 shows some rough estimates of the impact of biotechnology applications on output by the year 2005 in the EC.

There is, therefore, a prima facie case that technical advances leading to accelerated growth of production and labour productivity will be viewed with scepticism by policy-makers in many European countries, at least as far as domestic applications are concerned. These aspects are considered to have played a role in the EC restrictions on the production of isoglucose, a sugar substitute, and on the use of bovine growth hormones, though in the latter case consumer resistance and perceived health considerations are also important factors. A recent survey (OECD 1989a: 39) indicated that 'some industrialists mentioned bovine growth hormones as the most important example of a new biotechnology development mistake', because it offered the potential of increasing milk production and reducing employment in a sector already plagued by excess capacity. Further, to the extent that adoption of biotechnology innovations requires enhanced farmer capabilities for management, acquisition of know-how and investment, it will tend to strengthen the existing trends for an increasing proportion of total output to be produced by fewer and larger farms. This will tend to accentuate inequalities between farmers and, perhaps, regions. These considerations raise the wider issue (already discussed in Ch. 5) of how to reconcile the short- and medium-term adjustment concerns with the longer-term objective of achieving higher efficiency in resource use, which would not be served by restrictions in the application of cost-reducing technology on grounds other than ethical, health and environmental. If such technologies are progressively adopted, the mounting costs of providing market support for the increased output may well accelerate the pressure for reform of agricultural policies, including reforms that would allow an enhanced role for market forces.

It follows that the application of those advances in biotechnology having predominantly quantity effects will be conditioned by, among other things, the prospects for export of the technology itself, its products and/or the increased farm output. Additionally, remunerative domestic markets for increased agricultural output may eventually emerge from advances in biotechnology and evolving market conditions making for competitive production of non-food industrial products from biomass conversion, e.g. fuels and chemicals, although the prospects for major uses are presently rather remote (see Box 3.1 for ethanol; a thorough analysis of potentials is given in Lewis 1986). In the European context the biomass feedstock for such production could be provided by beet, wheat, maize, potatoes, sugar cane (in some Mediterranean countries) and Jerusalem artichokes, as well as by woody biomass from energy trees (poplar, willow, alder, eucalyptus, etc.).

Biotechnology advances with desirable effects on food safety, quality and the environment are likely to face better application prospects, although the subject is not devoid of controversy (e.g. release of genetically engineered micro-organisms for pest control) and the prospects are of long- rather than medium-term significance. On the food safety side, the major benefits may be forthcoming from an eventual reduced intensity in the use of agro-chemicals and of chemical residues in food, resulting from pest-resistant plants and from microbial biotechnology applications in plant nutrition and protection, including integrated pest control management. At the same time the development of pesticide-resistant crops may contribute to increased use of pesticides for weed control. The above mentioned OECD report (OECD 1989a) considers this to be a factor in the trends towards large-scale acquisition of seed companies by the agro-chemical industry. Here also biotechnology's contributions to diagnostics would make for more efficient detection of sources and prevention of food contamination. Environmental benefits from biotechnology are seen in the above mentioned longer term and still uncertain possibilities for reduced use of agro-chemicals and eventual selective substitution of synthetic by natural (biomass-based) products having improved biodegradability characteristics.

The policy issues raised by these scientific advances and their applications to food and agriculture involve aspects of ethics, safety, law and public information as well as more traditional economic ones. Questions of ethics and safety originate in the possible effects of biotechnology's potential to manipulate and modify living organisms. For ethics, the principal concern is the potential spillover effects into human genetics and eugenics, as well as issues of animal welfare and the treatment of animals as mere commodities. Safety considerations originate in the risk of possible unpredictable effects — and perhaps uncontrollable and irreversible ones — on health and the environment of the release of new organisms, in particular microbial ones. There are risks here, both environmental and economic, associated with the reduction in genetic diversity promoted by the wider adoption of plants and animals with uniform characteristics. While these areas of concern are legitimate and important, the policy debate has not been helped by the lack of clarity in public debate as to what biotechnology is and how to assess potential risks and benefits.[1]

Improved public information to demystify the subject (including highlighting biotechnology aspects which reduce risks in genetic research

[1]   Fleisher (1989), citing a number of authors, states that 'The scientific community has by no means reached a consensus on the degree of concern to associate with deliberate release of genetically engineered micro-organisms. Proponents of a "business as usual" approach argue that recombinant DNA technology is simply an extension of previous genetic manipulation technologies such as cross-breeding and hybridization. It is argued that the precision offered by the new technologies leads to a more predictable product and that concern should be diminished rather than enhanced by the use of recombinant DNA. Opponents of the "business as usual" approach argue that recombinant DNA leads to the mixing of genetic materials that would not be possible using traditional techniques and that this manipulation may lead to less genetically stable organisms. The idea that small genetic changes necessarily yield small ecological changes is not supportable, even though it is likely true in many cases. More specifically, the genetic manipulations have the aim of increasing virulence, shifting host ranges, and expanding the ecological range of organisms. These types of ecological shifts, if successful, are cause for close scrutiny of unanticipated risks to human health and property as well as the ecosystem.'

compared with more conventional breeding methods) and greatly strengthened institutional arrangements for safety assessment and risk management are among the policy responses to be considered. The excessive secretiveness surrounding the development of new technology is proving counterproductive. If adequate information is not made available, public opinion can be easy prey to influences promoted by various interests.

In the legal area the protection of new biotechnology products and processes is at issue, including through patents (for a survey see 'Copyright plants', *The Economist*, 27 September 1986). The subject is not new. For example, in the USA legal protection of asexually produced plants was introduced in 1930 (Plant Patent Act) and of those produced sexually (by seed) in 1970 (Plant Variety Protection Act), while in Europe the International Convention for the Protection of New Plant Varieties was adopted in 1961. However, the principle of granting protection under the general patent law has met with resistance, although the patenting of industrially useful micro-organisms has received wide acceptance in recent years. Traditional attitudes in law have been slow in evolving towards granting patent protection to genetically-modified plants and animals, except in the USA. The European Patent Convention excludes new plant and animal varieties. Among the major issues involved are how not to discourage R & D investment by industry while moderating the diffusion-restraining and cost-raising effects of patent claims, including claims to fundamental scientific inventions of basic research underpinning biotechnology applications. A related issue is the extent to which the increasing importance to agriculture of inputs subject to proprietary rights would increase the dependence of farming on the chemical and pharmaceutical industry. Eventual concentration of such proprietary rights in the hands of a small number of large firms and the existence of significant economies of scale, as for example is the case of BST production (Molnar and Kinnucan 1989: 4), would strengthen monopoly elements in the inputs industry which could then appropriate for itself part of the public support provided to farming.

The economic issues include the already-discussed mixed feelings of policy-makers towards a technology with potential to increase production and productivity in situations of already existing surplus capacity in agriculture. Naturally, countries of the region interested in accelerating production growth as well as the developing countries could benefit from the new opportunities if they have equitable access to them, other things being equal — in particular, risks in the health and environmental areas. Production and productivity increases in these countries could have longer-term trade impacts adverse to the agricultural exporting countries of the region. However, it is often the developing countries which lack the necessary infrastructure (scientific, technological, institutional) for technology development and diffusion. Profit-driven R + D by the private sector, typical of biotechnology, is unlikely to give priority to technologies without remunerative markets, no matter that such technologies could benefit large numbers of poor in the developing countries. On the contrary, it may happen that patent protection could work in the opposite direction, to the extent that it pre-empts further R & D and eventual cost-reducing innovations for products with mass market, but still unremunerative, potential. These considerations make evident the need for more emphasis on public sector and internationally funded R + D in biotechnology.

Constraints faced by the developing countries concerning access to the new technologies may be somewhat eased by advancing globalization, including the operation of multinational enterprises. Agricultural protection in Europe is also thought to encourage relocation of biotechnology firms to countries with cheaper biomass feedstocks (Lewis 1986). In practice, however, developing countries may be adversely affected by the new technologies if the industrialized countries use them to promote import, including uneconomic, substitution. The development of the HFCS sector in the USA, substituting essentially for imported sugar under border protection, is a case in point (see Ch. 3). The potential of producing cocoa butter by microbial process presents similar characteristics. The USSR's plans to expand production of single cell proteins for the feed sector is another case of biotechnology application that could have adverse trade impacts on oilseed exporters. The eventual entry of other oil and gas producers into this field using local feedstocks of low opportunity costs (e.g. flared gas) can have significant trade impacts for both developing and developed exporters of vegetable proteins for the feed sector.

Even genuine productivity-increasing technologies for certain export crops of the developing countries, e.g. cocoa, would benefit those countries and farmers best equipped to take advantage of such technologies to the detriment of the weaker countries and producers. When such productivity gains concern tropical products, particularly the non-competing ones, with main markets in the developed countries, much of the gains in productivity could be passed to the consumers in these latter countries in the form of lower prices if competitive market conditions prevail (for further discussion see Fowler *et al.* 1988; see also Box 12.1 in Ch. 12).

More generally, biotechnology may be viewed as another link in the technology development chain favouring shifts in comparative advantage to technologically developed countries and away from those possessing natural resources. In the same category of effects belong biotechnology's contribution to the strengthening of established trends towards declining raw material content per unit of GDP, although the above-mentioned potential of substitution of biomass-based products for synthetic ones would be a contribution in the opposite direction. In the area of relations with the developing countries belongs also the issue of their (and indeed worldwide) access to germplasm (whole plants, seeds, cell cultures and cuttings) — much of which originates in the developing countries themselves — so as to ensure that they share in the benefits, or defend themselves against adverse effects, of its exploitation. Among the policy responses to this concern at the international level is the adoption in 1983 of the International Undertaking on Plant Genetic Resources whose objective is to ensure that plant genetic resources of economic and/or social interest, particularly for agriculture, will be explored, preserved, evaluated and made available for plant breeding and scientific purposes (for issues related to the International Undertaking see FAO 1989).

Agricultural training and extension

Improved farmer education and training will be increasingly needed as the requirements for managing the farm business become more complex. In many countries the level of general education among the farm population compares unfavourably with that of the non-farm population, emphasizing the importance of a continuing education and training programmes for adult farmers. The vocational training of new entrants will also require increasing resources, particularly in the Southern European countries where farmer education tends to lag behind elsewhere. Some countries now insist that possession of a recognized agricultural qualification is a precondition for receiving state aid or even for entering farming itself.

The role of advisory and extension services, including those provided by the private sector in some countries, has been critical in the rapid diffusion and take-up of innovations by farmers. These services are provided in a variety of ways. In some countries agricultural extension is closely linked to agricultural universities and agricultural research and development centres, while in others it is organized by the farmers' own professional and vocational organizations. In the CPE countries where large-scale state and cooperative farms predominate, extension specialists are employed by these enterprises and their associations although close links are maintained with the agricultural universities and research centres.

The magnitude of public investment in agricultural extension is shown in Table 7.3. Although there are considerable differences in spending intensities between sub-regions, these are largely accounted for by differences in relative costs, and the range of manpower intensities is much narrower. In the future, increasing the quality rather than the quantity of the resources engaged in agricultural extension will be the priority task.

Table 7.3 Public extension expenditures and workers

Sub-region	Public sector agricultural expenditures as % of value of agricultural product			Extension workers per $10 million (constant 1980) agricultural product		
	1959	1970	1980	1959	1970	1980
Northern Europe	0.65	0.85	0.84	2.76	2.56	2.61
Central Europe	0.29	0.42	0.45	2.19	2.77	2.73
Southern Europe	0.11	0.35	0.28	2.00	2.76	2.69
Eastern Europe	0.32	0.36	0.40	2.36	2.88	3.13
USSR	0.28	0.32	0.35	2.26	2.33	2.50
North America	0.42	0.53	0.56	1.44	1.31	1.08

Source: Evenson 1986.

Studies of the economic return to extension show positive effects of extension programmes on farmer productivity. In the case of the USA, returns to investment in extension are well above the range of normal returns to investment. Studies show strong mutual reinforcing interaction effects between investment in research and investment in extension, although they also reveal an inverse relationship between extension programmes and farmers' schooling, meaning that an increase in the level of farmers' schooling makes extension services less valuable.

In many of the Mediterranean countries and in Eastern Europe and the USSR, increasing production will continue to be the main justification for extension programmes. Developments will emphasize the better management and improved professional training of the human resources engaged in extension and the fuller utilization of new information and communications tools. While these developments will also be important in North America and North-western Europe, extension services in these countries face the additional problem of a growing concentration of production in the hands of larger-scale specialized producers whose direct links with research stations render the role of generalist extension workers less relevant. Also on distributional grounds the allocation of substantial public advisory resources to working with such farmers can be questioned. Budgetary pressures have already led some countries (for example, the United Kingdom and Ireland) to introduce charges for advisory work, blurring the distinction between public extension workers and private consultants, and leading to fears that smaller and medium-sized farms with development potential may find it more difficult to get access to the advice and assistance they need. Public extension services may come to emphasize more issues such as community development and expand into non-agricultural areas such as small business development.

Conclusions

Agricultural research and technology will occupy an increasing place on the policy agenda of the study countries. The most important role for public authorities will continue to be to undertake worthwhile agricultural research and development of value to the society but not attractive to the private sector, and more generally to decide on research priorities. The regulatory role of national and international public authorities with respect to proprietary rights, diffusion and transfer of technological innovations will have to assume increasing importance. In addition, concern in research for ethical, health and safety issues will need to be kept under constant review, as well as concern for the impact of agricultural practices on the aesthetic qualities of both natural and artificial environments, and for the quality of life in rural communities. Finally, the impact of technological advances on international relations in agriculture, particularly those with the developing countries, is a factor that must be considered in policy formulation for research and technology.

8 Agricultural structure and rural development issues

Farm structure objectives

Agrarian structures reflect each country's particular historical heritage, resource endowments, and legal and traditional institutional arrangements, particularly with respect to inheritance and the functioning of labour markets. In the study countries the diversity of farm structures could hardly be greater, ranging from the fragmented holdings in Southern Europe where strong traditions favour retention of land in family hands from generation to generation, to the large, state-owned and collective production units characteristic of the CPEs.

Table 8.1 illustrates the distribution of farms by size in a number of European countries in the 1980s. The average landholding per farm has been increasing over time and is expected to continue to do so over the foreseeable future. In France, for example, the average landholding per farm grew at 2.7 percent p.a. over the period 1955–85 and expected growth over the next decade exceeds 3 percent p.a. (Tirel 1987). In the CPEs the average agricultural landholding per collective farm ranged from 1,900 ha in Romania to 6,400 ha in the USSR in 1985. The corresponding average figures for state farms were between 3,200 ha in Poland and 16,100 ha in the USSR. However, as noted below, the small family farm sector in Poland has remained predominant throughout the post-war period.

Any discussion of farm numbers and sizes and structural trends is importantly conditioned by definitions. Changes in the definition of what a farm is clearly influence the trend in the total number of farms, and the statistical practice of increasing the lower threshold for the definition of a farm (either in area or sales value terms) itself gives an upward bias over time to estimates of average farm size. In Poland, for example proposals are under discussion that the lowest limit for classifying holdings as farms be raised from 0.5 ha to 1.0 ha (Wos 1988). Measures of average farm size and the size distribution of farms also depend on the criterion of size used. If the yardstick for measuring farm size is income (or a proxy) per farm rather than hectares of land per farm, then the relative ranking of countries according to average farm size can change drastically because some hectares are more productive or are used more intensively than others. Thus, the average farm size of the UK at 69.3 ha is the largest in the EC while that of the Netherlands at 16.9 is just below the average of the EC, after Denmark, France, Ireland and the FRG (CEC 1989a); yet the Netherlands has the largest average farm size in the EC in terms of European Size Units (ESU, which measures the

Table 8.1 Distribution of number of farms and area for selected study countries, 1980s (in percent)

Country	1 ha – 5 ha 1980 (a)	(b)	1985 (a)	(b)	5 ha – 50 ha 1980 (a)	(b)	50 ha and over 1980 (a)	(b)	1985 (a)	(b)
Belgium	28.4	4.7	27.7	4.3	67.3	74.5	4.2	20.8	5.2	24.0
Denmark	11.1	1.2	2.0	0.2	78.8	59.8	10.1	39.0	15.4	43.5
Finland	30.8	7.9	...	...	67.8	83.2	1.3	8.9	...	...
France	20.6	2.1	18.5	1.7	66.1	54.5	13.3	43.3	16.5	49.5
Germany FR	34.5	5.4	30.0	4.6	61.6	74.4	3.9	20.1	5.3	24.5
Greece	72.0	39.1	69.7	30.4	27.8	57.0	0.2	3.9	0.7	16.1
Ireland	15.2	1.9	16.0	2.0	76.0	65.0	8.8	33.1	8.9	33.5
Italy (c)	68.4	21.6	67.3	19.2	29.8	47.9	1.7	30.5	2.1	32.9
Netherlands	24.0	4.1	24.2	3.8	73.0	81.0	2.9	14.9	3.8	17.7
Norway	50.2	16.1	...	...	49.7	83.9			...	...
Poland (d)	59.5	25.1	58.4	23.2	...	...	5.0	18.7	...	....
Portugal	77.9	18.7	...	...	20.3	26.0	1.8	55.3	...	...
Spain (e)	55.8	8.6	...	...	38.7	36.4	5.5	55.0	...	...
Sweden	16.3	2.5	...	...	72.4	55.1	11.3	42.4	...	...
Switzerland (f)	28.5	6.2	...	...	70.8	89.0	0.7	4.8	......	
United Kingdom	11.8	0.5	12.3	0.5	55.6	17.7	32.6	81.8	33.5	82.5

Sources: CEC (1989a) for the EC countries; Wos (1988) for Poland; OECD (1983) for the others.
(a)   Proportion of total number of holdings.
(b)   Proportion of total area.
(c)   1977 for the columns headed 1980.
(d)   Farm size limits: lowest class 0.5 ha–5 ha, upper over 15 ha, peasant farms only which account for some 76 percent of total area.
(e)   1982.
(f)   1969 for holdings proportions, 1975 for area proportions.

value of gross output less special variable costs, a concept akin to Gross Value Added, per farm). Similarly, the size distribution of farms within each country can look quite different depending on the criterion used to measure farm size. Use of an economic measure (ESU) for farm classification reveals the rapid process of concentration of production in the large (economic) – size farms that occurred in the EC countries in the ten years from 1975 to 1985 (Table 8.2).

Given the diversity, farm structure concerns differ widely between the study countries. In the CPEs the main policy thrust has been the creation of industrial conditions for rural labour through the formation of large scale production units, which, except in Poland (where private farms occupy 77 percent of agricultural land in use and account for 82 percent of net final agricultural output), are still the prevailing structure. After an initial phase of uniformity regarding the types of large-scale production units the picture is now much more varied. In Czechoslavakia, Hungary and Romania cooperatives play an important role, and in Hungary traditionally they

Table 8.2 Contribution of small and large farms to total output, in percent

| | Farms under 4 ESUs | | Farms over 40 ESUs | |
	1975	1985	1975	1985
Belgium	6.0	1.6	14.1	53.7
Denmark	3.3	0.6	16.7	60.2
France	6.5	1.9	20.8	51.5
Germany, FR	9.3	3.2	12.6	40.6
Ireland	26.5	8.9	4.8	22.5
Italy	32.5	12.0	19.1	43.1
Netherlands	2.1	0.3	22.5	77.0
Portugal	...	37.5	...	23.8
Spain	...	16.9	...	29.7
United Kingdom	4.0	0.7	46.5	81.5

Source: CEC (1989a).
Note: Farm size and total output measured in ESUs.

function with considerable autonomy. In Bulgaria state farms and agricultural cooperatives have become large-scale agro-industrial combines, with more or less the character of state enterprises. In all CPEs novel forms of associations and joint enterprises are being formed, with particular emphasis being placed on downstream links with the food processing industry. Further, private or small-scale agricultural production has existed and still exists in these countries. Since the beginning of the 1980s, in the majority of CPEs, measures have been taken to encourage the development of production in household plots and auxiliary farms. Not only in Poland, but also in Hungary and Bulgaria the role of small-scale production is especially important. In Hungary such production is treated as an integral part of the socialist system of agriculture; the larger enterprises organize and guide small-scale producers, ensuring the spread of up-to-date production methods and the sale of products (Csáki 1987). More generally, policy reforms under way in a number of CPE countries encourage this symbiotic relationship between socialized and private agriculture, and indeed such reforms envisage the promotion of quasi private farming by making it possible for persons or groups of persons to rent land on a long-term basis and farm it on own-account. Such encouragement has met with mixed success so far.[1]

Reforms affecting farm structures in the CPEs are currently in full evolution and one cannot be definitive as to final choices. A recent policy statement for Hungary reads as follows

We feel that the role of large holdings will be decisive but not exclusive. Farmers will be able to choose the forms of production ownership and organization and the size of holdings which best meet their needs. No form of ownership will have priority over any other kind of ownership. This will be guaranteed by State regulations. The

[1] 'Many peasants are reluctant to lease land or to keep a family farm. To a great extent, they became adapted to participation in the collective economic activity and prefer keeping up their households without changing the habitual ways of work and life' (Lukinov 1989).

reform of large ownership and land use is also something that we are looking at. Our goal is to ensure that citizens and commercial or producing organizations have the same chance, the same opportunities for use of land. The only principle that we want to maintain is the principle of economic management of the land. (Statement of the Hungarian Minister for Agriculture and Food to the 25th Session of the FAO Conference — November 1989, FAO document C 89/PV/10)

In most of Western Europe, structural policies were developed to guide and accelerate the process of labour outflow, to lower the proportion of non-viable farms and to create units that could produce at lowest cost. Policy measures aimed at easing farm operators out of agriculture by offering various incentives (pensions, training facilities) and using the land released to enlarge the remaining farms. Structural policy objectives have been less explicit in North America, where public intervention has been less important than in other countries in influencing the rate and direction of structural change. More recently, concern has increased that this spontaneous process is giving rise to a very concentrated farm structure which is threatening to undermine the survival of the family farm, and that the indirect effects of public intervention through price, tax and research policies have hastened this process. An additional factor has been the tendency for non-farm capital to seek investment opportunities in farmland. Long-term returns on farmland investment in the USA are estimated to have exceeded those in other investment areas, as follows:

Rates of return, 1960–88, percent p.a.
Farm real estate 9.67
Business real estate 8.81
Stock Exchange Index-Standard and Poor's 500 9.62
Long-term corporate bonds 6.57
US inflation rate 5.00
Source: Dunne 1989.

In this way part of the benefits of farm support programmes accrue to non-agricultural investors to the extent that they are incorporated into higher land values and rents.

Structural policies have also been influenced by considerations of agriculture's role in balanced regional development and the preservation of the rural environment. In many Western European countries policies to support agricultural activity specifically in mountainous regions or in regions of sparse population have been put in place. The EC, for example, supplemented its basic structural policy measures by a directive on farming in mountain and other less favoured farming areas in 1975 which provided for income aids to farms in these regions. Such aids comprise the payment of a compensatory allowance for natural handicaps. The number of farms qualifying for such aids increased from 340 thousand in 1975 to 1,050 thousand in 1987 (CEC 1989c). In North America rural policies focused on infrastructure development, including roads, electrification, postal and telephone services. Many of the relevant public institutions were created prior to the 1930s but are still in place, although their very success has in many cases led to their diminished importance. More recently, rural development projects in the United States received a new impetus as a result of the Rural Development Act of 1972 whose main purpose is to improve economic opportunities in rural areas through various financial aids. In Canada a coordinated effort has been made since the late 1960s to promote economic and social development in rural problem areas which are mainly situated in the more densely settled Eastern Provinces.

These structural measures, while softening the impact of adjustment for many farm families, have so far had little influence on the underlying trends. Although the farm labour force has fallen, a relatively large number of farm businesses remain too small to provide acceptable incomes to their occupiers. Rural development schemes have also had mixed success. Although rural depopulation has been halted or reversed in some rural regions, the demographic, infrastructural and economic structures of many others remain very weak. In a number of countries structural policies are being revised, partly in the light of these considerations, and partly in the light of the restricted opportunities for agricultural expansion. Some of the main directions of change are now considered.

Options in structural policy

The countries in Eastern Europe and the USSR will, in one way or another, have to address the issue of farm structures in the context of overall policy reforms. Some countries may opt for radical reconsideration of existing structures. The complex technical, economic and legal issues involved will have to be tackled taking into account a multitude of factors, such as historical experiences related to the evolution of present farm structures as well as overall economic and agro-ecological diversity. It is noted in this context that farm structure issues often transcend purely economic criteria because the legal relationship of persons and groups of persons to land, more than for other economic assets, and the associated income distribution patterns, have important implications for societal structures and attitudes. In examining alternative options for farm structures, consideration will be given to an increasing role for private ownership of land as well as to other arrangements appropriate for promoting a long-term and certain framework aimed at ensuring the independent management and market-oriented behaviour of farm enterprises. The experience of others is relevant here. Nearly all Western countries have directly productive activities in public sector ownership. Separation of management from ownership (whether public or private) is very common in the non-farm sectors, with various efficiency results that are not necessarily closely correlated to the form of ownership. It is also becoming of some importance in the farm sector in some countries. Obviously agriculture, more than other sectors, needs independent individual management. This proposition should, however, be carefully evaluated in the context of modern farming conditions and emerging trends in farm size.

In these countries novel forms of farm structure may have to be investigated in the reform process; these should meet the double criterion of independent management and market behaviour while, as far as possible, preserving objectives of social control of land resources in countries in which this is considered important for various reasons. In this context, note must be taken of trends in many countries which point to an increase in the average farm size and farm concentration. The future role of the large production unit which emerged under socialized farming should be evaluated in the context of such trends, taking account of successful cooperative experiences oriented towards the provision of services to production agriculture. Avoidance of land fragmentation will be an important consideration, though this is not equivalent to preservation of the prevailing very large production units, which probably suffer from diseconomies of scale. In all cases, the transition to more efficient farm structures will need to be actively supported through provision of services (e.g. research and extension, credit) for farms to become economically viable.

The approach to farm structures will also have to be seen in the wider long-term perspective of possible developments in the overall economy. In this context, account must be taken of the fact that in most countries in Eastern Europe and the USSR the share of agriculture in total employment continues to be high by international standards. Better overall economic performance will, in the longer term, be associated with expansion of employment opportunities in the non-agricutural sector. This would accelerate somewhat the outflow of labour from agriculture. The case is, therefore, strengthened for farm structures favouring an increasing farm size and further substitution of capital for labour. In parallel, account must be taken of trends everywhere in the industrial countries for part-time farming and multiple sources of income of farm households to play an increasing role in maintaining approximate parity between farm and non-farm incomes.

A central feature of structural improvement in many Western European countries has been the consolidation of fragmented holdings. Consolidation schemes often involve major works of water management, farm road construction, resettlement of farms and landscaping. In many cases there is no increase in the size of the consolidated holdings, the resulting structure is rather rigid and progress is slow and costly. While consolidation will remain important and necessary, budgetary pressures will lead to a search for simpler and less thoroughgoing arrangements in the future.

A second concern in most West European countries is to increase the rate of land mobility so as to bring about a larger number of viable farms more quickly. Many countries have introduced schemes of incentive payments to outgoing farmers conditional on the vacated land being used for structural improvement. Generally the impact of these schemes has been disappointing, possibly because the value of payments has been eroded by inflation over time and was not sufficient to compensate for the increasing value of land during the 1970s and the availability of other payments of a welfare nature to people who remain in farming. There is now a renewed interest in pre-pension or farmer retirement schemes in the context of taking land out of agricultural production altogether, and it remains to be seen if these schemes will attract a larger number of applicants.

The Southern European countries, considering the historical developments which have given them a very different farm structure, have followed a different path of structural reform. Large farms have been sub-divided and the land distributed among smaller holdings or used to create new units. These countries have also sought to ensure a more rational use of farm land and have taken measures to make abandoned or under-utilized land available to active operators through expropriation pro-cedures, high land taxes or incentives to rent out the land.

The changing age structure of farmers has prompted certain structural policy measures in some countries. In most countries there has been a steady increase in the average age of farmers (e.g. about half the farmers in France and Italy are over 55 and in Poland 56 percent of all farm heads are over 50), and many elderly farmers have no heir. Various measures can be taken, including concessions for the earlier transfer of land in the capital taxation code, exemptions for young farmers from some restrictions (e.g. quotas), the extension of pre-pension incentive payments to transfers within the family, and the inclusion of farmers in social security schemes, to try to redress this unbalanced age distribution. In addition, measures to facilitate entry into farming can be taken, including special establishment aids for younger farmers and the provision of tax relief on land transactions.

In Western Europe and North America the leasing of land has been seen as a way of facilitating entry into farming and encouraging greater land mobility. Tenant farming contributes to the former objective by separating the capital requirements for land ownership from the requirements for operating capital, while it can contribute to the latter objective by permitting outgoers to retain ownership of their land while transferring its use to another farmer. However, the extent of tenant farming has been gradually declining in the region in response to legislative attempts to provide greater security to tenant farmers and to control rents. It could be rehabilitated by, for example, a clearer legislative framework which avoids giving excessive protection to the tenant farmer and by removing discriminatory taxation provisions which discourage renting. A factor to be taken into account here is that the benefits of agricultural policies, which are meant to support the incomes of farmers, will partly accrue to non-farmer landowners via higher rents and land prices. In some countries public landholding has been used to create a 'farming ladder'. For example, the Province of Saskatchewan in Canada has over half a million acres of land in a purchase and lease-back programme primarily designed to help young farmers enter agriculture. Leasing of land, while retaining ownership in state hands, is also likely to attract greater interest in CPE countries. This process has already started in the USSR and in some other countries although the pace of implementation varies.

In some areas the creation of a network of viable farms through farm amalgamation or leasing is not a practical proposition. Part-time farming and dual jobholding is a possible alternative strategy to maintain and increase farm household income. The extent to which this development is welcome or not will depend on a government's weighting of land use and farm income objectives, but the trend is towards an increasing acceptance of the positive role of part-time farming in contributing towards the economic and social stability of depressed farming areas. In the CPE countries, too, one may foresee an expansion of non-agricultural employment and a growing role for non-agricultural income in total household income. This will take place in a number of ways: more pluri-activity among families employed in state and cooperative farms, the development of agricultural production activity by non-farm families on private plots, and more diversification by large-scale farms themselves into non-agricultural activities in order to better utilize available labour and to increase total incomes.

With the growing evidence of the concentration of farmland into fewer and fewer hands structural policy in Western Europe and North America will increasingly turn to address the problem of maintaining a sufficient number of 'family farms'. For example, a recent study of the US Office of Technology Assessment (US Congress, OTA 1986) suggested that the number of farms in the USA could decline from 2.2 million in 1982 to 1.3 million by the year 2000, with the decline concentrated in the small and moderate size classes (the number of large size farms was expected to increase). Another study (Commins and Higgins 1987) sees a 20 percent fall in the number of farms in the EC-9 between 1983 and 2000, with again the decline concentrated in the small and moderate size classes. In a number of countries a trend towards a dualistic agricultural structure has been noted, with a large number of increasingly marginalized farms on the one hand and an ever-smaller number of very large farm businesses on the other. Concern is expressed about the social and political consequences of this trend and the perception that the family farm institution is under threat, undermined further by the growing power of large, oligopolistic firms in many agricultural markets.

It is probable that reports of the demise of the family farm in North America and Western Europe have been exaggerated. However, government policies, particularly price, taxation and technical innovation policies, have contributed to a faster rate of polarization than would otherwise have occurred. The impact of possible reforms in these areas on structural evolution are difficult to predict. Lower prices may hasten the exodus of smaller farms and thus contribute to an improvement in farm structure in the long run, but in the short run they will exacerbate the lack of viability of smaller farm units. Quantitative restrictions, such as marketing quotas or set-aside arrangements, will not necessarily affect the pace of structural change if quota rights are transferable when land changes hands. Specific structural measures will continue to have the largest impact. These include fiscal reform to reduce the pressures for farm enlargement and to encourage the break-up of very large units, environmental regulations which restrict intensive livestock production on a large-scale, and direct quantitative ceilings on farm size. Such measures would also make a positive contribution to redressing market imbalances.

Rural development policies

Agricultural activity is becoming less important in rural areas throughout the study countries. Rural area problems are no longer coterminous with agricultural sector problems, and rural development is no longer synonymous with agricultural development. This trend is clearly shown by comparing the proportion of total employment in agriculture with the proportion of the total population living in rural areas. For example, even in the more agrarian countries of the EC (Ireland, Italy, Spain) the percentage employed in agriculture is less than half the rural population. In the EC-12 out of a total of 116 regions there are only 10 (in Greece, Italy, Spain) in which farming still accounts for 30 percent of employment generally (CEC 1988f). Thus governments concerned with balanced regional development will increasingly turn to a wider range of policy instruments in their attempt to reduce regional disparities in income and living standards.

Rural development is a field in which the objectives of Western and Eastern Europe are progressively converging. In, for example, Hungary and the USSR the role of non-agricultural activities in the rural economy is being increasingly emphasized. Indeed, a number of state and collective farms report that favourable income trends are more due to non-agricultural activities (e.g. food processing, brewing, feed milling, wholesale and retail sale of foodstuffs) than to agricultural production. In Czechoslovakia collective farms in 1986 were making over 50 percent of their total profits from 'subsidiary' (i.e. non-agricultural) production. In Hungary, the integration of agricultural and non-agricultural activities (industry, construction) in the context of farm enterprises often ensured the economic viability of agricultural production (by increasing capacity utilization of capital and labour resources and providing investment funds) and contributed to rural development and to the more balanced spatial distribution of industrial activities (Marton 1988). Increasingly, the non-agricultural activities of these units are not at all connected with the agro-food complex. This trend to greater diversification of the rural economy is likely to strengthen further.

However, rural areas and their problems are not homogeneous within countries, and are even more divergent across countries, as a result of, *inter alia*, differences in resource endowments, remoteness and historical patterns of settlement and land use. Rural areas in many countries of North-western Europe with high densities of population are frequently close to major urban centres, and agriculture accounts for a relatively small proportion of their economic activity (rural areas Type I in Figure 8.1). Production and marketing conditions are generally favourable, and jobs outside of farming are easily available. Such urbanized rural regions in both Europe and America experienced significant population growth in the 1970s and first half of the 1980s. In some cases the substantial demand for land for new housing and for recreational opportunities created conflicts with agricultural interests. Policy responses in such cases must be directed at controlling risks of damage to the rural environment, rather than encouraging economic activities indiscriminately. Physical planning and environmental policies have an important role to play in resolving such conflicts.

Other rural regions are influenced by problems of peripherality such as poor demographic structure due to loss of population in the past, inadequate off-farm employment opportunities, poor infrastructure and a low level of service provision. In these regions the main problem is usually to diversify

Figure 8.1   Policies for agriculture in rural development in the European Community

	RURAL AREA TYPE		
	I	II	III
Type of problem	Pressure of demand for land. Land fragmentation. Various types of pollution.	Structural handicaps. Migration to medium-sized towns. Abandonment of marginal land.	Poor agricultural structures. Difficult production conditions. Difficult living conditions.
Type of solution	Land-use planning. Protection of the environment.	Individual measures to improve structures (production, marketing). Quality policy. Diversification. Guidelines for accompanying measures for market policy. Nature conservation.	Maintenance for farmers to stay on their farms. Nature conservation.

Type of scheme	Demarcation of agricultural areas (land-use plans). Land consolidation; keeping agriculture competitive; aid measures to encourage the changeover to forms of agriculture which use fewer chemicals. Extensification (see under Type II).	Improvement of structures. Directives on labels and designation of origin. Organizational support. Aids to farmers who undertake to pursue environmental protection objectives. Identification of coastal areas suitable for aquaculture.	Income support. Improvement of the rural habitat. Soil conservation. Compensatory allowances (differentiated). Creation of protection areas. Premiums for environmental maintenance.

Source: Material extracted from CEC, 'The Future of Rural Society, Commission Communication to Parliament and the Council', *Bulletin of the European Communities*, Supplement 4/88.

Note: Rural Area *Type I:* areas near to or easily accessible from the big conurbations, particularly in the centre-north of the EC and in many coastal areas; *Type II:* areas of rural decline particularly in the outlying Mediterranean parts of the EC; *Type III:* Areas furthest from the mainstream of EC life, with difficult access, such as mountain areas and certain islands.

the economic base of the area away from dependence on low-productivity agriculture to industrial and service activities. Public investment to improve local infrastructure and to upgrade human resources through education and training, and provision of investment incentives to attract footloose manufacturing and service employment are common policy measures. Advances in information and telecommunication technologies hold out the eventual promise that work locations will be less physically restricted than at the present, but such technologies will probably not have major implications for the attractiveness of rural locations for manufacturing and service activities within the time frame of this study. Of major importance, within the EC context, is the introduction in 1985 of the Integrated Mediterranean Programmes for France, Italy and Greece to enable them to adapt as smoothly as possible to the new situation resulting from the accession of Spain and Portugal. Also in 1985, a special 10-year programme was launched to bring Portuguese agriculture in line with the economic realities of Community membership. Institutional innovation through setting-up broadly-based development boards or agencies with the brief to promote and coordinate development activity in the region has also been undertaken.

The Single European Act of the EC gave new impetus to 'The goal of stepping up measures to increase economic and social cohesion and, more specifically, to narrow the gap between the regions of the Community, particularly by the promotion of rural development' (CEC 1989c). To the extent that such objectives are pursued by means of agricultural policies, including policies for strengthening agricultural structures, they must be in harmony with the wider objective of reform towards a more market-oriented agriculture. Recent EC policy proposals for rural development envisage that policies for agriculture (shown in Figure 8.1) would be only one component of total interventions and they would be tailored to the particular characteristics of the different rural areas.

In CPE countries the large-scale production unit structure and the associated integration of agricultural and non-agricultural activities provides a potentially fruitful source of entrepreneurship and development skills which can be used to diversify the local economic base. Such measures achieved a degree of success in the 1970s and in many remoter rural regions population decline has slowed down or even been reversed. However, significant gaps between rural and urban areas in social infrastructure and services persist in some CPEs, as indeed in other countries, particularly those of Southern Europe. In this context the near equality of nominal wages in agriculture and other sectors of the economy reported in the statistics for some CPEs (see Ch. 11) would not ensure near equality in living standards if many goods and services were scarcer, and by implication more expensive, in rural than in urban areas (for a description in rural/urban discrepancies in USSR see Morozov 1987:72). Recent policy initiatives emphasize measures to correct such imbalances.

Conclusions

In both the centrally-planned and market-economy countries of the Study there is at present considerable development in both structural policy objectives and instruments. In the CPE countries, except Poland, the fundamental structural transformation has been completed with the creation of large-scale production units, and in some countries attention now focuses on ways of exploiting the strengths of these units through novel forms of management of the enterprises and of association with each other, with downstream industries and with the still-growing-in-importance private plots and auxiliary farms. But, as noted, the issue what are the policy options in the area of farm structures for a more efficient agriculture continues to be central to the policy reform debate in many CPE countries and radical changes may be in store. In the western countries, where the maintenance of the family farm structure is a durable policy objective, there is a perceptible trend away from direct intervention to consolidate and amalgamate smaller holdings into viable units based on agricultural production alone, and a greater acceptance of the permanence, and desirability, of pluri-activity. Concern for the family farm now focuses on the threat from the increasing concentration of production on larger farms and on the possible contribution of public policies to this trend.

In a broader context, a second tendency to be observed in both East and West is a shift away from sole or even main emphasis on agricultural production increases as the route to rural development and balanced regional growth, towards greater recognition of the need to diversify the economic base of rural areas and to consider the problems of public service provision to a largely non-agricultural rural population. This change has been encouraged both by the limited overall scope for increased agricultural production, and by the realization that the intensification of agricultural production in disadvantaged rural areas has often involved the substitution of capital for labour and further encouragement to labour out-migration. But while integrated rural development has become a popular catchword, its efficacy in reversing rural decline remains to be proven.

9 Resource and environmental management issues

Resource and environmental objectives

Problems of resource availability and conservation will require much more attention from governments in the region. There is a growing awareness that environmental conservation cannot be left as an afterthought but rather should be an objective ranking in importance with agriculture and forestry. Since the countryside lends itself to multiple uses — food, fibre and timber production, recreation and tourism — there is a prima facie case for the adoption of an integrated policy approach in which agricultural, forestry, environmental and amenity objectives are pursued simultaneously. The debate is often centred on the adverse environmental effects of intensive agriculture. It is, however, equally important to recognize the positive effects of agricultural activity on the environment, particularly in certain regions, e.g. the Alpine countries where agriculture contributes to control flooding, soil erosion and avalanches and to the preservation of the rural environment (see FAO/ERC 1988 for a general introduction to agriculture–environment issues in Europe).

Agricultural and non-agricultural developments have increased the pressure on the agricultural resource base, and at times raised some concerns about the future adequacy of resources such as land, water and energy for agricultural production. At the same time, there is increasing criticism of the effects of intensive agricultural production on other environmental resources (see Figure 9.1). However, environmental pressures differ between the study countries. In the countries of North-western Europe, the pollution of surface and groundwater supplies by fertilizer and slurry run-off, the deterioration of landscape quality due to the intensification of farming, the destruction of wildlife habitat and the problem of land abandonment in marginal farming areas are among the issues requiring attention. In general, it should be noted that the agriculture/ environment interactions and the need for policy integration are of somewhat different nature and urgency in relatively densely populated and intensively farmed Europe than in North America. In this latter region the use of fertilizer is comparatively less intensive on the average and proximity of farming operations (particularly livestock farming) to urban areas less of a problem.

In many countries, particularly in Southern Europe, the USSR and in parts of North America, soil erosion due to over-exploitation is now increasingly recognized as a widespread environmental problem. Soil productivity loss is only part of the damage caused by soil erosion. Off-site damages, resulting from sediment deposition in water bodies and pollution, are usually a multiple of on-farm productivity loss. In Turkey over 50 percent of the agricultural land is threatened by erosion. In the USSR much of the area in the 'Virgin Lands' is affected by erosion and is believed to be marginal. In Southern Europe, the pursuit of regional development objectives entails the risk that the introduction of new agricultural technology and/or other activities, e.g. tourism, will disturb some of the last remaining natural habitats of Europe. Environmental problems related to agricultural practices have also become a primary concern in Eastern Europe and the USSR, a region which also faces serious environmental problems of non-agricultural origin. In these countries soil erosion, declining soil fertility, soil compaction due to the use of heavy agricultural machinery which is favoured by the large state and cooperative farms, soil acidification from both fertilizer use practices and aerial pollution, deterioration of ground water quality, and the salination of irrigated land threaten the success of planned production intensification and even the maintenance of existing production. For example, irrigated crop production (e.g. cotton) in the Aral Sea basin in the USSR may be adversely affected by the adoption of urgently needed water conservation to protect the lake from further deterioration (the lake's surface is estimated to have shrunk by 40 percent since 1960 resulting in 26,000 square kilometres of salty desert). In Hungary it is estimated that the first water supplying layer is polluted all over the country and its water is not suitable for direct human consumption. In the same region the negative environmental effects of heavy concentrations of livestock (especially pigs) on large industrial units are becoming increasingly evident.

It would be difficult to develop homogeneous policy stances and to design a set of pan-European measures which would adequately reflect this sub-regional, country and local diversity of environmental characteristics and issues. Beyond environmental diversity, countries also differ in the degree of public awareness and policy stances on environmental issues. Such diversity reflects in part differences in the seriousness of environmental problems. For example, the policy debate and the search for solutions is most advanced in countries facing serious environmental problems, including the Scandinavian countries, the Netherlands and the FRG. Other explanatory variables of inter-country differences seem to include relative income levels and the associated development objectives, perhaps because the most prosperous countries attach higher values to environmental goods and services (relative to other goods and services), e.g. clean air and water, or amenity services of the countryside. For example, in the Netherlands 'the environment has soared ahead of unemployment as the prime national concern; some 70 percent of the Dutch would forego higher living standards for a cleaner country' (Clark and Dixon, *Financial Times*, 12 April 1989).[1]

---

[1]   It is worth quoting the authors' comment on the differences in policy stances on the environment within Western Europe: 'The split corresponds so nearly to Europe's oldest fault-lines — Teuton/Latin, Protestant/Catholic, bourgeois/agrarian — that one could devise elaborate cultural theories to account for it. Dutch, German and Scandinavian officials offer a simpler explanation: their countries (whether because of dense populations or proximity to the appalling pollution of Eastern Europe) are simply the worst affected'.

Figure 9.1: Effects and Consequences of Intensive Cropping Practices

Intensive agricultural practices	Pollution and Contamination				Natural Resource Degration					
	Soil	Water	Air	Food	Soil quality	Water/acquatic resources	Forest resources	Landscape amenities	Biotope and habitat disturbance	Loss of wildlife and genetic diversity
Drainage of wetlands, land reclamation	Accelerated pollution due to loss of "ecosystem services"	Accelerated pollution due to loss of "ecosystem services"			Soil dehydration, degradation	Changes in ground water table, water cycles		Loss of natural areas values for conservation	Negative effects on water-related ecosystems; Loss of terrestrial and aquatic habitats	Loss/extinction of flora and fauna species, particularly fish and water fowl
Conversion of pastures, forests, etc.					Reduction in nutrients; changes in soil hydrology; erosion	Flooding, siltation sedimentation of water systems	Loss of forest vegetation, ground cover; Loss of pasture for game/domestic animals	Decrease in recreational amenities; Loss of heritage values; degradation of rural landscapes	Loss of complex biotopes of special ecological value, e.g., natural forests; destruction of habitats	Loss/extinction of species diminished variety of wild life; Loss of genetic resources
Consolidation of fields, removal of hedges, walls, etc.					Inadequate management leading to degradation	Negative effects on water conservation and management		As above	Reduction in number and complexity of habitats maintained by the traditional agricultural ecosystems	Loss of species abundance and diversity, particularly birds and insects
Tillage, use of heavy machinery			Combustion gases		Compaction of soil; wind and water erosion; reduced productivity	Heavy silting, sedimentation of water systems		Landscape degradation	Disturbance of soil ecosystems	Adverse effects on soil organisms, microflora and microfauna

Activity										
Application of synthetic fertilizers	Nitrogen saturation; concentration of heavy metals	Nitrates, heavy metals leached into surface and ground waters	Evaporation	Nitrate contamination of food crops, shellfish	Reduced fertility; accumulation of heavy metals	Eutrophication; contamination of aquifers	Loss of alluvial forests	Loss of recreational amenities	Destruction of biotopes and loss of terrestrial and aquatic habitats due to pollution, contamination and eutrophication	Loss/extinction of wide range of species; diminished diversity of plant and animal life; adverse effects on soil organisms, etc., loss of genetic resources
Spreading of manure, slurry	Nitrogen saturation; concentration of phosphates, heavy metals	Nitrates, phosphates, heavy metals leached into surface and ground waters	Release of amonia leading to acidification	Nitrate contamination of foods crops, shellfish	Reduced fertility; acidification; structural damage	Eutrophication; contamination of aquifers	Damage from acidication	Loss of recreational amenities, aesthetic value	As above	As above
Application of pesticides	Residues, degradation products	Residues, etc. leached into surface and ground waters	Evaporation; spray drift	Residues in food crops, livestock	Accumulation of residues	Contamination of Aquifers	Loss alluvial forests	Loss of recreational amenities	As above	Loss/extinction of non-target flora and fauna, particularly fish, birds and insects; build-up of resistance in plants and insects
Irrigation	Excess salts	Salinisation; accelerated pollution			Salinisation; water logging; erosion	Saline contamination of aquifers; reduction in ground water levels		Landscape degradation	Negative effects on aquatic ecosystems, habitats	Loss/extinction of species; reduction in ecologist diversity
Straw-burning			Local pollution					Reduction in local landscape amenities and aesthetic values	Disturbance of biotopes, habitats; risk of destruction by spreading fire	Reduction in species

Source: Reproduced from FAO/ERC (1990).

In the more specific area of agriculture explanatory variables of inter-country differences would include the relative weight of the sector in the pursuit of development objectives, the degree of self-sufficiency already attained, the existence of surplus capacity and the associated contradiction inherent in costly farm support policies which maintain or increase such capacity and have adverse environmental effects. There are some interventions (e.g. the creation of nature reserves, soil conservation measures) which countries can pursue independently. It will be, however, difficult for countries to employ controls unilaterally which will leave them at a competitive disadvantage — unless in response to strong pressures from environmental lobbies — but indeed some countries are doing so (e.g. the Netherlands, Denmark, Sweden). Given the exigencies of trading in international markets, multilateral initiatives will be easier to adopt also because interdependence of agriculture and the environment often cuts across national frontiers and likewise benefits of the integration of environmental aspects into agricultural policies may accrue to different countries. It is sometimes the case that a country could obtain higher environmental benefits from investment in pollution abatement in a neighbouring country than in its own territory.

The search for policy solutions in this area will most appropriately involve extension to inter-country relations of the policy analysis theory and practice concerned with the provision and payment for public goods and services, i.e. those for which it is infeasible or undesirable to establish property rights and markets, such as clean air (with the consequence that individual consumers cannot be charged according to the amounts they use, nor can they be excluded), and when consumption by one person does not reduce the amount available to others. It will also involve similar extension of policy analysis in the presence of externalities, i.e. when a country's actions impose costs (or confer benefits) on another country, for which the latter is not compensated (or does not pay) through the market mechanism. An apt example is river pollution from farm and industrial activities in Country A adversely affecting downstream Country B (for discussion of public goods and externalities see Harvey and Whitby 1988). The application of these principles in the formulation of multilateral policy responses requires some degree of political cohesion and sense of common destiny of the European body politic. Supranational policy initiatives like the ones already implemented and increasingly taking place in the context of the EC, e.g. adoption of safeguards for environmentally sensitive areas in the context of the EC socio-structural policy, the imposition of common clean water standards, etc., are significant steps in the right direction. They contribute to the removal of obstacles emanating from the above-mentioned national differences in public perceptions, relative national development priorities and economic and administrative capabilities to undertake the necessary action.

Ensuring resource availability

In some countries, the maintenance of the size and productivity of the land

and water base can be a major issue. In the USA concerns are voiced that greater production is being achieved at the expense of excessive soil erosion, over-use of ground and surface water, and irreversible loss of wetlands and prairies (USDA 1988a). In densely populated countries competition from other land uses, particularly urban use and roads, and from forestry and recreation uses in more marginal areas, is leading to a gradual reduction in the land available for agricultural production within the European region. It is difficult, however, to assess to what extent these land use changes should be a major cause of concern, since continuation of technological change should compensate adequately for such losses. The case for non-intervention is favoured by those who believe that market decisions can best bring about the optimum allocation of land among its alternative uses. However, there are serious doubts whether the market for land adequately reflects the long-term need for food production. The irreversibility of many major changes in land use is a persuasive argument that social interests are not necessarily reflected in land market decisions.

Changes in the taxation of agricultural land and zoning are among the policy options to retain agricultural land. Where land is subject to property taxation, changes in tax legislation can be an effective way of slowing the transfer of farm land to non-agricultural use. In the USA, for example, most states have shifted from market value assessment to use value assessment in the case of farmland in the expectation that the lower property tax burden will encourage farmers to keep their land in agricultural use even in the presence of incentives for conversion. Capital gains taxation designed to appropriate a significant proportion of the gains from changes in land use can play a similar role. Alternatively, land use zoning can be enforced. This approach requires substantial resources to introduce and may meet with resistance from those who object to the constraints imposed and to the changes in land value which zoning may entail.

In countries where irrigated agriculture is important the availability of water will be a source of concern. Often, water development projects are heavily subsidized and water is supplied to agriculture at below cost. Without these subsidies, a number of farming enterprises in some areas would not be viable. However, uneconomic rates of offtake have been encouraged and as surface water usage approaches its maximum potential and pressure on groundwater lowers watertables, major shortages or major adjustments in agriculture are to be expected. The policy debate will focus on whether the resource should be allocated via the price mechanism, entailing a rise in user fees, or by direct regulation; whether where water rights are appropriated, a rationing scheme should be adopted and/or certain farming practices prohibited. Related to this is the issue of water allocation between agriculture and competing industrial and urban uses, particularly in water-poor areas. Integration of agricultural and environmental policies should take account of the double effects of irrigated agriculture: it accentuates water shortages in areas where this is an issue and encourages more intensive use of agro-chemicals. For example, it is estimated that in the USA irrigated farms use more than twice the fertilizer and three times the pesticides of non-irrigated farms (USDA 1988a: 26).

Protecting environmental resources

The intensification of agricultural production has led to a number of threats to the environment. Pollution problems such as nitrate contamination of groundwater supplies, eutrophication of surface water supplies and marine waters and smell, have arisen from the increased use of fertilizer and problems of waste disposal from large livestock units. Conservation fears have been aroused as the number and diversity of flora and fauna are reduced by the practices of hedgerow removal, wetland drainage, monoculture and increased fertilizer use. The amenity value of the countryside may also be reduced as varied and intimate landscapes are replaced by prairie-like fields, and as traditional *de facto* rights of access are ploughed under. In some countries soil erosion and, a related problem, the abandonment of agricultural land are major issues.

Given the link between agricultural pollution and the growing intensity of agricultural production, one approach receiving increasing attention to reducing agricultural pollution is to return to lower-intensity farming by reducing the density of animal agriculture, by reducing the use of agricultural chemicals, or by shifting to less intensive land uses. There is considerable debate over how these goals should be achieved. A shift to agricultural practices using less of the environmentally harmful inputs would not be economically viable if it only meant reverting to the less-polluting farming practices of earlier decades, except when premium prices are paid for food grown without chemicals. It is more a question of combining knowledge from traditional practices with that of modern science and management. In economic terms this would require substitution of labour, particularly skilled labour, and management as well as of land (e.g. through crop rotation) for agrochemicals. The fact that such substitution is not occurring spontaneously indicates that the economic fundamentals do not favour it. If, as most evidence seems to indicate, current rates of agrochemicals use reflect profit maximizing behaviour on the part of the producers (see, for example Daberkow and Reichelderfer 1988), the mere existence of alternative technology for low-input agriculture will not lead to its adoption. For this objective to be achieved some change in key economic or social parameters is required (e.g. prices of inputs and outputs), as well as more research and the removal of structural impediments to the diffusion and adoption of such technology (e.g. through training and extension).

Given that lower-intensity farming would also contribute to the production goal of redressing market imbalances, many of the arguments echo those considered in Chapter 5, e.g. reduction in support prices, taxes or quotas on inputs, etc. In evaluating policy changes for their impact on market balance and on the environment, it is important to review existing support policies for bias in favour of input-intensive technology. Beyond the general impact of such policies in stimulating production above market requirements, and hence higher use of agrochemicals, there are also specific policy characteristics which influence crop mixes towards more input-intensive crops and/or tend to 'freeze' cropping patterns and practices in favour of monocultures and against diversification into patterns with crop rotations. The US commodity programmes are thought to have such characteristics because (a) they provide a disincentive to reduce acreage devoted to a supported crop as this may reduce the farmer's base acreage on which future programme benefits will be calculated, and (b) they link eligibility to receive support for a given crop to the cross-compliance condition of non-increase of base acreage devoted to any other crop which is also supported. Thus, for example, shifting acreage from nitrogen-intensive maize (140 lb/acre in 1985) to less intensive soyabeans (15 lb/acre) is severely penalized. The same policies may also stimulate use of agrochemicals to maximize yields beyond the level dictated by current conditions (market prices plus support payments) on the expectation that higher current yields will increase the base for receiving support payments in the future. Although this positive link between current yields and expectations of future deficiency payments was severed in recent revisions of support policies, it is thought that some farmers still operate on the expectation that the link will be re-established in the future (for more discussion see National Research Council 1989: 69–85, and Young 1988).

If direct income payments were introduced to compensate farmers for income falls following lower prices and other measures restricting production, the environmental impact could be strengthened by linking their payment to the achievement of environmental goals, e.g. through management agreements with the beneficiary farmers. The EC has moved in this direction by introducing the possibility of making payments to farmers for the conversion and extensification of their holdings. A land set-aside programme without provisions for environmental activities would have a relatively insignificant effect on agricultural pollution, and could exacerbate it as farmers farm their remaining land in production more intensively than before. The taxation of production inputs, in particular fertilizer, could significantly alleviate the pollution problems caused by this input provided the taxation was levied at a sufficiently penal rate, although the environmental benefits will accrue with a considerable time-lag (20–30 years).

It is generally accepted, at least in the OECD countries since 1972, that the 'polluter-pays principle' is a sound basis for policy analysis in deciding how to go about protecting the environment. The principle implies that pollution control costs should be reflected in the production costs and prices of the goods and services whose production causes the pollution. This would be achieved by making the producer pay for such costs. The producer would then pass on the costs in the form of higher prices, though this possibility depends on market characteristics, as discussed below. It is obvious, therefore, that although the (polluting) producer may be the 'paying agent', costs will ultimately be diffused throughout the economy, including upstream and downstream sectors of agriculture, domestic final consumers and eventually foreign importers (for net importing countries foreign suppliers may gain). How the total cost will be shared out has to do with demand and supply elasticities with respect to price. Given relatively low demand elasticities for primary farm products it is likely that the consumers rather than the producers will bear most of the cost, since higher product prices will result in increased consumer expenditure and less in reduced quantities consumed, although the declining share of farm value in total final expenditures on food will tend to blunt this effect.

There are cases when agricultural producers as a whole will gain rather than lose from internalization of anti-pollution costs following application of the polluter-pays principle. This may happen when the loss in producer surplus of the farmers who will use less of the polluting inputs will be more than counterbalanced by gains in such surplus (following higher product prices) of producers of the same and substitute products who were using little of the polluting input in the first place, e.g. because of agro-climatic differences. This could be the case in large unified markets with regional diversity in which any given farm product and its close substitutes are produced under widely differing input combinations. On this point a recent USDA study (Barse *et al.* 1988) found that the banning for environmental reasons of all soil fumigants used to control soil-borne pests in the USA for citrus, potatoes, tomatoes and tobacco would result in a net gain to all producers of these crops of $380–600 million and a net loss to consumers of $3.0–5.1 billion. The net gain to all producers is made up of $100–200 million losses of the fumigant-using producers and $480–600 million net gains of non-using producers. The range of the estimates reflects alternative estimates of price elasticities of demand.

USDA pesticide studies for other crops report similar findings. For example, removal in the USA from the market of corn and soybean pesticides with alleged environmental and safety risks could increase US agricultural production costs, crop prices, farm incomes and consumer expenditures, causing farmers to gain and consumers to lose. Some farmers may lose but farmers as a whole would gain. Consumer losses, including those of the downstream industry, would be higher than the farmers' gains (Osteen and Kuchler 1986). Similarly, Osteen and Suguiyama (1988) estimate that removal from the market of chlordimeform, a cotton insecticide, implies farmer gains of $197–691 million and consumer losses of $345–1500 million (the higher estimate assumes increased insect resistance to complementary insecticides — pyrethroids).

The above findings provide some preliminary insights useful for analysing policy options. In the first place, aggregate farm income (often an important policy objective) need not be adversely affected following controls of pesticide use. Secondly, the income distribution effects within agriculture may favour small farmers (another important policy objective) if non-users or less intensive users are also small farmers, although this need not be so and more research is needed to find out. Thirdly, farmer income maintenance or increase reflects rising market prices consequent upon pesticide controls. It will, therefore, occur only if producer prices are permitted to rise and, indeed, rise above intervention prices or those used for determining deficiency payments. If those market and income support measures are active, the first beneficiary of rising prices will be the public budget. Fourthly, the controls considered in these studies result in a net welfare loss to society in the sense that farmer (and eventually the budget) gains are lower than consumer losses.[1] This net welfare loss should be set against the health and environmental benefits resulting from such controls. This is a more apt indicator compared with simpler ones of public budget costs for environmental protection.

Some important qualifications apply to these findings, however. As noted, in the usual case when producer prices are supported either through border measures and intervention buying as in the EC, or through deficiency payments as in the USA, the introduction of cost-increasing environmental regulations would not necessarily lead to increased producer prices. In such cases, i.e. when there is no competitive equilibrium in the markets before the introduction of environmental regulations, producers would suffer a loss (producing less at higher costs but receiving the same support prices, hence reduced net transfers from taxpayers and/or consumers, assuming higher costs do not induce policy-makers to increase support prices). Government expenditure would be reduced and consumers may or may not pay higher prices (US and EC support systems, respectively). Consumers may even gain on the aggregate if taxes are reduced as a result of budget savings (for analysis and empirical estimates for the USA see Lichtenberg and Zilberman 1986). The other important qualification is that in the presence of unchanged border measures, increased imports may prevent domestic prices from rising or, in the case of exporting countries, exports will be reduced. In both cases producers would lose from environmental regulations while consumer prices may not rise appreciably, assuming the country has no dominant position in world markets and international prices do not increase. Such effects may create pressures for policy changes (e.g. import restrictions or export subsidies) to counteract the effects of environmental regulations. This is an important aspect which must be given due consideration in the efforts to achieve multilaterally agreed reform of agricultural support and protection policies in the context of the Uruguay Round of Multilateral Trade Negotiations (see Ch. 12).

These findings of the US pesticides studies may also apply to eventual restrictions on the use of mineral fertilizer or manure to control nitrate pollution, a central problem in Europe and parts of North America. The issue here is rarely total ban, except perhaps in special water catchment zones, but rather reduced use and improved methods of application. Reasoning by analogy is probably not appropriate, given differences in market organization and the role of prices in the USA and Europe. The literature on the subject (e.g. OECD 1989b; HMSO 1988) usually refers to farm income losses and the need for compensation, consequent upon restrictions on nitrate use, though the reference is usually to the sub-set of

---

[1] The fact that higher production costs are reflected in higher consumer prices and not necessarily in reduced total farm income is in broad agreement with the time-honoured tradition that the gains from technological change in agriculture are passed to consumers or, where prices are supported, to the owners of land. Banning or restricting certain productivity-raising inputs is in practice a reversal of traditional technological change. For an analysis of the effects of shifts in the supply curve on producers' surplus see Miller *et al.* (1988).

affected farmers, not total farm income.[1] To the extent that input-use restrictions for environmental purposes are considered only for relatively small areas, e.g. nitrate protection zones, the resulting decline in aggregate agricultural production would be too small to influence average prices. In such cases farmers as a whole would not be compensated through the market in the form of higher prices, with the result that affected farmers may have to be compensated by other means.

While general measures to reduce the intensity of agricultural production would contribute to environmental objectives, they will need to be supplemented by specific measures to encourage the more sensitive management of agricultural inputs and wastes, which even under a lower-intensity agricultural regime will continue to exist. The optimal approach is to design a series of monetary incentives and disincentives to implement the polluter-pays principle. Where such price-based incentive systems prove difficult to implement or administer, a regulatory approach might be adopted under which farmers would be required to follow approved practices and fined for not doing so. Application of the polluter-pays principle is problematic when the polluter cannot be readily identified, e.g. in the case of non-point pollution. Even when the polluting practices are known, the most cost-effective policy response need not always involve restriction or outright discontinuation of such practices. A UK study on the nitrate problem (HMSO 1988) indicates that the most cost-effective solution would generally involve a combination of land use controls and water treatment.

Various approaches are also open to resolve the various conflicts between agriculture and conservation and amenity interests. Gaining the voluntary cooperation of landowners and operators, and providing advice and training in environmental management should always be part of the effort, as should the public acquisition of sites considered worthy of protection, to form, say, national parks or nature reserves. But those measures may not be enough. Some form of development or planning control is probably a necessary instrument of environmental management, but its limitations as a conservation policy must be recognized. It will meet political objections and will prove difficult to enforce. It can cause real hardship to farmers in environmentally sensitive areas, and it may not be an effective way of maintaining the resource in question. If the payment of compensation is required when a particular farm development is prohibited, then the financial implications will seriously limit its effectiveness. On the question of compensation, it should be noted that the income losses of affected farmers may overstate the cost to society of environmental protection. In the typical European country with producer prices maintained above world market prices, the social value of forgone output (net of production costs saved) will be less than its valuation at domestic prices (for estimates and discussion see Willis *et al.* 1988).

---

[1]   OECD (1989b: 78) notes, however, that 'the introduction of nitrate quotas as a means of reducing agricultural surpluses should not be discounted and in areas where pollution is a problem they should be seriously evaluated because of their favourable impact on farm income'.

A number of countries are considering the use of, and some are actually implementing, positive incentives for conservation. Such incentives are usually offered in the form of management agreements in which, when agreements are primarily restrictive, payment is the means of offering farmers compensation for the loss of potential income, or, otherwise, is a return for the maintenance or production of valued resources or services. However, as long as the use of land for agricultural production is so heavily subsidized, then alternative uses of land such as conservation or amenity purposes must also be heavily subsidized in order to make them competitive.

Another area which will require increasing attention in the future is the effect of industrial pollution on agriculture. The effects of nuclear contamination arising from the Chernobyl accident brought home these dangers in a stark manner, as well as the fact that many of these effects do not recognize national boundaries. Similarly, acid deposition from the burning of fossil fuels as well as the improper disposal of sewage sludge and industrial effluents have adverse effects on agriculture, forestry and fisheries, both locally and across national frontiers. Studies on the economic effects of acid deposition and ozone increases indicate (paradoxically, though only apparently so) that farmers may gain from increases in atmospheric pollution (Crocker 1989). The argument is analogous to that reported earlier in connection with the US pesticide studies. Increased pollution reduces yields of some crops and increases market prices in the presence of low demand elasticities. Farmers in the aggregate gain and consumers lose by more. The effect is a net social loss as conventionally measured in economic terms without counting the health and other benefits from a less-polluted environment. These results are subject to the qualifications mentioned earlier concerning the role of foreign trade and the operation of support and protection policies.

Soil erosion is now recognized as a significant environmental problem in many countries, although the extent of the problem is usually very unevenly distributed, and its costs to agriculture are not yet well established. Farm support policies in general and those which link benefit eligibility to acreage farmed in particular, e.g. some US commodity programmes, provide incentives for cultivation of more land than otherwise would be the case, including marginal and erodible land. Environmental damage occurs both on-site (soil erosion and compaction, loss of organic matter, threat to sustainability of production) and off-site (sediment deposition and associated pollution causing shortened reservoir life, clogging of waterways, loss of recreational values, increased costs of dredging and protection of water quality, etc.). Crosson (1989) reports that if present rates of cropland erosion in the USA continued for 100 years, presumably not accounting for the benefits of the current US conservation reserve programme — see below, crop yields at the end of the period will be 5–10 percent lower than they would otherwise be. Off-site damages are generally estimated to be significantly higher than on-site ones, 4 to 12 times as high according to USDA estimates (USDA 1988a: 15). Integration of agricultural and environmental policies would, therefore, require that proper estimation of off-site damages be assigned priority in evaluating the costs and benefits of agricultural policies.

A number of approaches have been developed for responding to soil erosion problems. Some indication of the future direction of soil conservation programmes, at least in the market regimes found in many countries in the Study, can be gleaned from recent US decisions. The 1985 US Food Security Act establishes a long term conservation reserve of up to 45 million acres with payments made to farmers who take highly erodible and other environmentally critical land out of production for 10 years and agree to implement an approved conservation plan (land to be under grass or tree cover and not to be used for hay production or grazing). The emphasis is on inducing farmers to voluntarily adopt soil conservation measures, generally by providing technical assistance and economic incentives. The latter include imposition of cross-compliance conditions, i.e. eligibility for benefits from agricultural support programmes is made contingent on the adoption of soil conservation practices and prohibition of converting highly erodible land (sodbuster) and wetland (swampbuster) to cropland. In the latter case, the implied penalty from the loss of programme benefits for non-compliance has been estimated at $48/acre (Ervin and Dicks, 1988). The US conservation reserve programme met with significant success with over 19 million acres having been enrolled by April 1987, more than the programme had anticipated. Further increases are reported after that date, with the total enrolled area having reached some 30 million acres by mid-1989 (for more details and accomplishments to 1987 see Dicks *et al.* 1988, and USDA 1988a).

In Southern Europe one of the main causes of soil erosion is deforestation, including through forest fires, and overgrazing. Policy responses in, for example, Portugal and Turkey, emphasize reforestation and pasture development. In Eastern Europe and the USSR, investment for land improvement must be increasingly directed to either safeguarding or restoring the fertility of the better agricultural land and not to the cultivation of marginal lands, unless ways of better utilizing the natural grassland resources of mountainous regions can be found. Additionally, a system of contracts with farmers, under which they undertake to implement specific soil conservation practices in return for government payments, might be developed. Another approach is regulation, requiring farmers to follow specific cultural practices or outlawing others, but regulation is generally regarded as a rather extreme solution and so far is not very widespread.

It is worth noting that different approaches to limiting production growth (Ch. 5) will have different implications for the soil conservation objective. If lower prices are effective in reducing production, they will encourage less intensive farming on all soils, including the more marginal ones. In a set-aside programme, however, land is withdrawn across all farm types and regions and not necessarily on farms where the land is most erosive. The overlap between land set-aside programmes and land diversion or withdrawal for conservation purposes must, however, be emphasized. According to the OECD report (OECD 1989b: 76) 'in the USA, experience is revealing that it is much easier to achieve care and maintenance by calling the set-aside land a conservation reserve'. It should also be noted that this overlap entails the risk that the resolve to pursue land withdrawal for conservation purposes, particularly if it is not tied to a long-term programme, may be weakened when market conditions improve and the need for production controls and for cost savings is accordingly reduced.

In a number of countries there are concerns about the effects of abandoning land on the ecological balance. Often agricultural decline occurs first in areas characterized by hilly topography and steep slopes where the continuation of farming required constant maintenance. On abandonment such areas may take time to regain their natural vegetation, giving rise to problems of soil erosion and an increased incidence of forest fires or avalanches. However, one EC study which evaluated the effects of land reversion on erosion, climate, water balance, flora and fauna as well as on countryside appearance and recreation, found both positive as well as negative environmental impacts (CEC 1980). It concluded that programmes to assist agriculture in marginal farming areas were largely justified on social rather than environmental grounds. This is not equivalent to saying that it is environmentally safe to leave land withdrawn from production in bare fallow, particularly in the non-marginal areas of North Europe. Increased nitrate leaching and loss of amenity value may result. The EC land set-aside scheme allows leaving land fallow but obliges the participating farmer to ensure that it is properly maintained with a view to protecting the environment.

Conclusions

Management of the environmental impacts, both positive and negative, of agricultural production will be an increasingly important challenge for the study countries in the next decade. It is argued here that, because the countryside lends itself to multiple uses, an integrated policy approach in which agricultural, forestry, environmental and amenity objectives are pursued simultaneously is required. This in turn may require institutional changes in the study countries to reflect the interdependence of activities and objectives which heretofore have largely been considered separately. Long-range land use and water basin planning will become more important.

There may well be differences in the pace at which countries will wish to pursue environmental objectives, reflecting different evaluations of environmental services and the seriousness of environmental problems on the one hand, and the conflicting claims on limited resources on the other. This could be a source of potential problems in international trade in the future if different environmental standards proliferate and act as non-tariff barriers to trade. On the other hand, if countries with tougher environmental standards eschew border controls or do not provide compensatory export subsidies, they will find themselves under increasing competitive disadvantage. Multilateral initiatives will be essential to resolve these issues in the future, as well as to encourage policy integration of agriculture and the environment in cases when mutual influences are exerted across national boundaries and the potential benefits from integration may accrue to different countries.

10 Agricultural marketing and food policy issues

Introduction

In all the study countries agricultural policy is increasingly seen as one component of an overall food policy. Food policies in the past have often been narrowly seen as simply concerned with ensuring basic food security, which in turn was often interpreted as the ability to meet a high proportion of domestic food needs from own production. Today, the risk of widespread food shortage in the study countries must be considered remote. Instead, food policy has widened to embrace issues such as food price stability, nutrition and health, and food safety and quality.

Agricultural marketing

In all the study countries a tendency towards the closer integration of the various activities which make up the food chain — input supply, primary agricultural production, food processing and distribution — can be observed. On the one hand, a higher proportion of food undergoes some form of processing, and the percentage of consumer expenditure on food which returns to the farmer is falling (Ch. 3). On the other hand, traditional marketing channels are shortening as producers and processors sell directly to large retail stores and the role of intermediate wholesale and auction markets declines. A third form of integration increasingly observed is ownership integration, as input suppliers develop forward linkages into primary production (feed compounders and animal agriculture), processors develop backward linkages (meat slaughterers and feedlots, vegetable canners and horticultural production) and retail supermarkets either establish their own food manufacturing plants or effectively 'rent' production capacity through the close specification of 'own label' brands. This process is encouraged by ownership integration promoted by multinational enterprises in food processing and distribution. For example, ownership of foreign hypermarkets by French companies increased from 1 in 1969 to 59 in 1983 (Traill 1988). In the CPE countries similar trends are evident in greater cooperation and closer association between enterprises in the agro-food complex.

Although food processing is not the primary focus of this study, aspects of these trends have implications for the development of primary agriculture and the achievement of agricultural policy objectives. In particular, farmers are being made increasingly aware of the dangers of remaining purely commodity producers, i.e. producing a commodity and then looking for the best price for it, and of the importance instead of adopting a marketing approach, i.e. identifying what the market needs and then producing to those standards and quality specifications. As commodity markets decline in importance, so do the prices obtainable on these markets relative to prices achieved for direct sales. Food commodities which do not reach acceptable quality standards will be sold at a substantial discount. In some countries this marketing orientation has been pursued through the development of formal contracts. This has been taken furthest for vegetable, pigs, poultry and egg production where timeliness of operation, continuity of supply and consistency of standards are of most importance. In the EC official policy has encouraged the formation of producer groups and associations whose function is to strengthen their market bargaining power, improve the quality of products and to even out the seasonal flow of marketings. In parallel, the remaining monopolistic powers of Marketing Boards continue to be eroded in the process of creating a single European market. Developments in some CPEs indicate a similar trend, e.g. Hungary, Poland, the USSR, where large-scale production units have developed processing and marketing facilities and also cooperate with individual plots in the surrounding area.

The closer integration observable in the food chain has been accompanied, particularly in the market economies, by substantial shifts in the relative power of different agents in the chain. The continued trend towards concentration in the input supply, food processing and food retailing sectors has increased the concerns about non-competitive behaviour at the expense of the primary producer. Governments need to consider whether existing measures to provide countervailing power to producers in markets where they buy and sell are adequate in the light of these trends. Existing examples of policy measures of this type include the right often granted to producers to market jointly their products, thus exempting them from antitrust legislation, and the more favourable tax treatment often given to agricultural cooperatives to encourage farmers themselves to become involved in downstream and upstream activities.

Food prices

Because of the wide differences between the study countries in the proportion of total household expenditure spent on food, the objective of food price stability assumes a different importance in different countries. Retail price stability has been pursued most vigorously in the CPE countries, at least until the mid-1970s when the burden of retail price subsidies for the state budgets became exceedingly heavy. Food prices are not usually a live political issue in the high-income countries in the study as long as food prices are increasing no faster than the general rate of inflation. In the 1970s increased variability in commodity markets contributed to greater variability in food prices in some of these countries, so that on occasion food prices began to lead the overall rate of inflation, which itself increased over the period. Because of the visibility of food price increases due to the frequency of food purchases and their role in the inflation psychology of the public, they became an important focus of policy-makers' concern.

Continuing sensitivity to inflation will ensure that food prices remain an item on the policy agenda.

Governments wishing to control food price increases have the option of reducing support prices for farm products, although the declining share of farm product value in the retail price means that their impact on food prices becomes less and less significant. Import restrictions can be relaxed to increase supplies to achieve the same purpose. Alternatively, policies which try to reduce marketing costs by stimulating competition and improving efficiency in the food marketing system may be preferred. Retail outlets might be required to give more price information to shoppers, antitrust legislation can be used to limit the concentration of retail outlets in particular catchment areas, and powers to control directly either food prices or food marketing margins can be used.

Subsidies on foods will also moderate food price increases, but at a high budget cost, as the recent experience in the CPEs has shown. Food subsidies accounted in recent years for nearly 14 percent of the entire state budget in the USSR and about 13 percent in the German DR (to take the two extreme examples). The pursuit of retail food price stability becomes increasingly costly when producer prices and food consumption are rising. Under these pressures several countries, e.g. Bulgaria, Hungary, Poland and Romania, have tried to ease the subsidy burden by allowing food prices to rise. More recent changes in Poland liberalized all food prices except for a limited number of basic items. As already mentioned, in the USSR preparations are being made for comprehensive price reform. It is foreseen, however, that state prices for basic foods will be maintained unchanged for another two to three years (Ambassade de l'URSS — Berne 1989). The speed at which the reduction or removal of food subsidies can be implemented is constrained by the relatively high share of consumer expenditure going to food in the CPE countries and the immediate welfare loss to consumers, while compensating gains from the more efficient allocation of resources take longer to materialize. In addition, the abolition or reduction of subsidies will call for greater outlays to old age pensioners and generally low income families, particularly those with children, will alter the cost calculation in institutional feeding programmes and require the supply of a greater variety of non-food items in the marketplace.

The question of compensation for income losses following food price increases receives increasing attention in the policy debate. Aganbegyan (1988: 182), in discussing food price reforms in the USSR, is emphatic on this point: 'the standard of living of the population as a result of the reforms should not be allowed to fall. Those who proposed price increases also proposed the introduction of rises in wages, pensions and other income, so that families would be fully compensated for the possible price rise.' It is, however, difficult to see how negative effects, on at least some population groups, can be avoided entirely. This is being increasingly recognized in the relevant economic debate. The key issue here is how to formulate and sequence such policies so as to create the conditions required for resumed growth as soon as possible, while safeguarding the living standards of the most vulnerable population groups. In this context, food price reform would still be beneficial, even if it only meant that the budget funds saved would be entirely used to subsidize the incomes of the persons affected, assuming this is administratively possible. The change from price to income subsidies would help to remove distortions and contribute to the transition to some form of market-clearing price mechanism in the whole economy.

The issue of consumer food prices will continue to be at the centre of the reform debate. Some countries in Eastern Europe, e.g. Poland, have opted for nearly complete liberalization of food prices at an early stage. The USSR, on the other hand, has opted for maintaining state prices for basic foods unchanged while preparations are being made for a comprehensive price reform. The early indications are that real hardship did occur in Poland for the weaker population groups while in the USSR the food situation has not improved. Among the policy responses contemplated, and occasionally implemented, for the transition period is Romania's temporary moratorium on food exports and, in some countries, resort to increased food imports, including access to imports at concessional terms. Increasing resort to food imports of countries with severe balance of payments problems involves yet wider issues of international cooperation in support of the reforms.

Nutrition and health

Although overall food availability is more than adequate in all countries in the region, problems of undernutrition associated with poverty, lack of knowledge and lack of access to food continue to exist. The elimination of malnutrition remains a major policy goal. Low-income households, one-parent families and elderly people are particularly at risk. The continuous influx of immigrant labour from the developing countries into some countries of Europe adds a new group at the bottom of the income distribution scale, which is most likely to face problems of access to adequate food.

Food assistance programmes can take a variety of forms. Outside the CPEs, the subsidization of major foodstuffs is not widely practised because of the large leakage of benefits to families not in need and thus the high cost of the programme in terms of its impact on malnutrition. Programmes are preferably targeted on groups considered particularly at risk. They can take the form of direct food distribution (school lunch programmes, meals-on-wheels for the elderly), food stamp programmes or general welfare programmes, as well as the provision of dietary information.

In most CPEs apparent per caput food consumption levels are relatively high, encouraged by improved supplies (even at the cost of substantial imports in some countries) and by subsidies on food retail prices. (See, however, Chapter 3 for reservations concerning the reliability of the Food Balance Sheet data in some countries.) Staple foods account for a large share of the diet; per caput consumption of meat, fruits, vegetables and high-quality processed foods is relatively low in most of these countries compared with most other country groups of Europe. More diversification in diets and ready availability of better quality foods are increasingly demanded. However, the constraints on reform of the food subsidy programme, mentioned above, make adjustments in food policy interventions difficult. Indeed, heavy food price subsidies, more restricted import policies and promotion of food exports in some cases, together with problems of general economic growth in the early 1980s, necessitated rationing of certain food products in some countries.

In a number of countries the policy focus has shifted to the health problems arising from overnutrition and eating the wrong foods. Many studies have linked diet to the increasing incidence of the so-called 'diseases of affluence' such as heart disease and strokes. A number of countries have formulated food and nutrition policy guidelines which recommend that people cut down on the consumption of fats, calories, sugar and salt as positive steps to improved health. Yet food consumption trends have often been moving in a contrary direction, though there is recent evidence that some changes are now occurring. Traill (1988) reports that in the EC countries 'the limited available evidence suggests that consumers have adjusted their consumption patterns in a manner consistent with current nutritional advice'. Nutritional guidelines have often been controversial, although the epidemiological evidence and the results of direct intervention studies provide sufficient justification for such guidelines. There is now consensus among health workers in Europe and elsewhere that reduction in the consumption of saturated fat, sugar and salt are beneficial for health and for the prevention of certain forms of diseases of affluence.

Views differ on the appropriate role for governments in attempting to influence what people eat. Programmes designed to give consumers more information about the nutrient composition of the foods they eat (nutrition labelling) and more knowledge on how to make use of this information (nutrition education programmes) have been broadly acceptable. More interventionist approaches, such as the use of fiscal policy or attempting to influence the composition of the food supply through price and trade policy have not been widely adopted in Western countries. To the extent that these programmes are successful, imbalances on some agricultural markets, particularly red meat, sugar and dairy products, will be exacerbated although demand for other agricultural products, such as poultry meat and fruit and vegetables, will increase. It must also be noted that current efforts towards trade liberalization could lead to lower prices and would tend to favour consumption of some currently highly protected commodities (e.g. livestock products, sugar) whose consumption should rather be reduced than increased on nutritional grounds. In some CPE countries the food subsidies and trade policies tend to influence consumption patterns in directions which are sometimes at variance with nutritional considerations. This is particularly true in cases of subsidies favouring consumption of livestock products, even when relatively high levels of consumption have been attained. Restrictions in food imports which may affect access to fresh fruit and vegetables belong to the same category of policies with adverse nutritional effects.

The impact of demand trends favouring healthier and better quality food on the food processing industry can be profound. For example, it is foreseen that in the European Community the subsectors of fresh/chilled and frozen foods may increase their shares in food consumption by 100 percent and 25 percent respectively and that of canned food may shrink by 25 percent by the year 2000 (Traill 1988). The food processing industry is responding to some extent to the challenge of healthier eating with the introduction of foods containing fewer of the undesirable nutrients (e.g. sugar, fat) or questionable additives, though it also contributes to increasing the share of foods with undesirable nutritional characteristics (see below). In general, the food manufacturing and preservation sector is better equipped than agriculture to adapt its processes and product mix to changes in demand trends, because of continuous technological development, including in the field of biotechnology (see Ch. 7). The rapid globalization of the food processing and marketing sectors through the activities of multinational enterprises contributes greatly to this process. The impact is not, however, always in the direction of better quality and healthier food consumption, e.g. promotion of soft drinks — often at the expense of the market share of natural fruit juices — expansion of 'junk' food consumption, etc. In particular, the industry recycles a large amount of surplus butter and uses salt as a taste enhancer in some processed products. With the increasing share of processed and fast foods in total consumption, these practices contribute to increase the amounts of fat and salt in the diet. The process of globalization may also contribute to decrease variety and diversity in food consumption habits, including regional diversity through its adverse effects on the number of small rural-based food firms.

Food safety and quality

The safety of the food supply has long been considered a responsibility of governments. The perishability of food and its ability to carry bacteria and transmit disease has led to public regulation and inspection programes which have been remarkably successful in limiting these risks. In recent years, issues of food safety and quality have broadened to include: food hygiene and sanitation; farm production practices, including the use of pesticides, antibiotics and hormones, and animal production conditions; and food processing practices, including the use of artificial ingredients, flavourings, colouring, preservatives and irradiation.

Policies in the food safety area are often developed in response to consumer group pressures. It is important, therefore, for the public to be correctly informed as to food hazards, since erroneous perceptions are not uncommon. For example, it is reported (Traill 1988) that food additives and pesticide residues are ranked by consumers as representing the highest food risks, followed by en-vironmental contaminants, nutritional imbalance, microbial contamination and natural toxicants. By contrast, scientists rank these risks in almost reverse order, as follows (in order of decreasing importance): microbial contamination, nutritional imbalance, en-vironmental contaminants, natural toxicants, pesticide residues, food additives. It is also reported that 'US studies indicate that the dangers from microbe infections are at least a hundred times greater than those from additives and pesticide residues' (Bloom, 1989). Food health and quality aspects will continue to be a live policy issue of increasing importance. The response should include continuous research into health threats, improved transparency of guidelines and standards for enhancing consumer confidence, including into the ability to monitor compliance and suppression of fraud. Traill (1988) states that 'to the best of existing knowledge (and with a considerable margin of error) accepted treatments are safe when properly adhered to. The types of scandals which undermine confidence in the industry as a whole are almost always the result of deliberate fraud.'

If consumer groups are successful in setting more stringent standards or in insisting on zero tolerance rather than the concept of acceptable risk, then some farm production practices which lead to higher physical productivity may be prohibited. Since some competitive advantage on world markets would thus be lost, unilateral action on this front will be resisted. Widespread prohibitions, on the other hand, would slow the rate of productivity improvement in agriculture, alleviating the short-run problem of market imbalance in some parts of the region but, if not strictly necessary on health grounds, imposing costs on society in the longer run.

Improved food quality, which is most often associated with higher food prices, has implications for the food subsidy programmes in CPEs. Since consumers are likely to accept higher retail prices if better quality food is supplied, maintaining stable food prices, which severely strains food subsidy budgets, is unnecessary. On the other hand, since production costs are also higher, the subsidy element could remain more or less unchanged. Indeed, as Hungary and Poland have found, when retail prices are rising, it becomes all the more difficult to abolish subsidies.

Conclusions

The increasing importance of a wider range of food policy concerns in the study countries broadens the number of constraints which must be taken into account in achieving agricultural policy goals. In some cases food and agricultural policy goals operate in tandem. For example, higher retail food prices in CPE countries, in conjunction with a broader supply of non-food commodities, could lower budgetary costs, raise incentives to producers and free funds for investment in food processing and distribution, thus reducing waste and improving food quality. Similarly, the prohibition under consumer pressure of certain practices associated with intensive livestock production could contribute to redressing market imbalances in market economy countries. In other cases food policy and agricultural policy objectives may conflict. For example, greater awareness of the health implications of certain patterns of food consumption may switch demand away from certain products and exacerbate over-supply problems on these markets.

None the less, under the influence of more active consumer lobbies, and in recognition of the growing links within the food chain, agricultural ministers are taking an increasing interest in food policy matters. There is a growing awareness that the basis of a prosperous agriculture is supplying people with what they want, and that farmers must pay much closer attention to changing demands in the marketplace. Open-ended price support guarantees in the past have tended to insulate farmers from these market pressures and have encouraged a commodity production orientation. The need to encourage a greater marketing orientation for agricultural production will require greater flexibility in the way such price support guarantees work in future.

11 Returns to farm resources

Farm income objectives

Maintenance, improvement and stabilization of farm incomes figure prominently among the objectives of agricultural policy in all countries; yet there is considerable ambiguity as to how this objective is accounted for in actual policy formulation. Consequently, prediction and *ex post facto* evaluation of the impact of policies on farm incomes becomes difficult. Part of the problem lies in the greatly diverse measures used to define the policy objective and in the difficulties in obtaining the appropriate data. Practices vary among countries. Sometimes it is the net income of the farm operator from the sale and own-consumption of agricultural produce, sometimes it is the value added in agriculture per person classified as employed in agriculture, whether or not a farm operator, and in the CPE countries it is sometimes the average wage per employed person in agriculture. Differences between the value added and net farm income figures (the latter excludes wages paid to hired workers, interest paid on borrowed capital and rent paid for land) can be substantial. For example, in 1985 total net farm income was 29 percent of net value added in agriculture in the UK; in the USA the comparable figure was 44 percent. The picture is complicated by the fact that incomes of the farm population from non-farm sources are significant and there are also non-farm people receiving incomes (rent, wages) from agriculture. Further, in any evaluation of living standards, access to public services such as health services, schooling and income maintenance schemes should be taken into account.

Despite the deficiencies of standard national accounts data on agriculture (agriculture GDP, NDP or NMP) as measures of farmers' income, they are often used to make comparisons between countries over time because they are the indicators most widely available. The farm income objective is usually defined in terms of per caput incomes in agriculture in relation to those in the economy as a whole (as, for example, in OECD 1987). It is important, however, to keep in mind that national accounts data measure relative labour productivities and they are not very useful for measuring living standard differentials among sectors. The data in Table 2.1 bring out the wide differences among countries in the relative productivity of labour in agriculture. In Table 11.1 the evolution of this variable over time is presented and compared whenever data permit with alternative indicators of relative incomes. It is important, however, to emphasize that many of the available data refer to average farm incomes, usually as percent of the average income of the economy as a whole. This figure is largely meaningless for policy analysis which is equally concerned with the income

Table 11.1 Relative productivities and incomes in agriculture, * selected countries

	GDP per person employed in agric. % of same in whole economy			Other indicators of relative income per caput		
	EUROSTAT data					
	1970	1977	1985	1970	1980	1983
1. ˙EUR 7 †	50	51	46			
(Germany,FR)	(39)	(44)	(31)	76(1972)	68(1982)	
2. Ireland	58	82	67			
3. Greece	...	54	67			
4. Portugal	...	37	33			
5. Spain	37	43	35			
	OECD Data					
	1968	1977	1985			
EC-12	42	42	40			
6. Finland	47	58	63			
7. Iceland	...	76	77			
8. Norway	42	54	43			
9. Sweden	53	64	60			
10. Austria	44	45	40			
11. Turkey	41	42	31	USDA data		
12. Canada	49	61	60			
13. USA	53	76	68	70	80	69
	ECE data			CMEA data		
	1971–75	1976–80	1981–5	1970	1980	1985
14. Bulgaria	84	72	58	96	86	86
15. Czechosl.	50	41	44	93	97	101
16. German DR	78	68	61	...	...	...
17. Hungary	88	75	77	99	96	96
18. Poland	57	44	51	84	100	100
19. Romania	55	54	47	93	96	97
20. USSR	61	46	46	83	88	96

Sources: Eurostat, *National Accounts ESA, Detailed Tables by Branch 1987; OECD, Historical Statistics 1960–85* (shares of agricultural value added in total GDP divided by shares of employment in agriculture in total civilian employment); ECE data (ratios of shares of agriculture in material sphere total NMP and employment), UN/ECE (1987, 1988); Germany (FR), 'Einkommensverteilung nach Haushaltsgruppen', *Wirtschaft und Statistik*, 7/1984, Table 2 (per caput disposable income from all sources of farm households % of that of all households). USDA, *Economic Indicators of the Farm Sector, Income and Balance Sheet Statistics 1983*, ECIFS 3-3 (per caput personal disposable income from all sources of the farm population percent of same of total population); CMEA data from *Statisticheskii ezhegodnik stran-chlenov SEV 1978, 1986* (CMEA Statistical Yearbooks, 1978 and 1986), Moscow (average monthly wage of workers and employees in agriculture, percent of that in the whole economy).
*   See also footnotes to Table 2.1.
†   EUR 7: Belgium, Denmark, France, Germany FR, Italy, Netherlands, UK.

positions of the different groups of farmers. In typical situations, a large number of small farmers (40–50 percent of the total in the UK and USA) produce only a tiny share of total farm output and the top 10–15 percent of the farms account for the bulk of production. In such situations it is difficult to see how a general farm income objective can be defined and pursued, particularly by means of policies which encourage production.

The farm income of farm households as measured using national accounts data, however, no longer provides a good indicator of the economic well-being of farm people, who frequently have off-farm jobs that generate substantial income and may receive 'unearned' income in the form of social security payments, investment income, and remittances from family members working away from home. The data on average wages per employed person in agriculture and forestry relative to the average national wage in CPE countries shown in Table 11.1 indicate that in most countries a steady improvement occurred in the 1970s and virtual equivalence has now been achieved. In these countries many non-farm activities organized by the large-scale units in agriculture are counted as part of agricultural sector activity and hence influence the reported income statistics. Farm workers in these countries enjoy equal conditions of social and legal security in such matters as pensions, sickness benefits, working hours, leave of absence, working relations and so on. Still, the near equality of nominal wages earned in agriculture to those earned on the average in the whole economy may not ensure corresponding near equality in living standards if, as happens in some countries, many goods and services on which incomes can be spent are much less available and, by implication, much more expensive in rural areas (see Ch. 8).

For countries in the region for which separate data are available, off-farm income has increased substantially over time, relative to farm income. Dual jobholding and part-time farming, once considered merely a halfway step to outmigration, have become an established feature of the agricultural sector. The wide country differences in the proportion of farmers with 'other main gainful employment' was already noted in Chapter 2 (38 percent in the FRG, 9 percent in Denmark). More systematic information is presented in Table 11.2. Country differences in the gap between the numbers working on agricultural holdings and the labour input they supply to agriculture (in terms of full-time equivalent workers) are significant. For example, in Italy and Greece it took 2.4 persons working on agricultural holdings to provide the labour equivalent to that of a full-time worker. In Denmark, the Netherlands and the UK the corresponding figure was 1.3. These differences are indicative of differences in the existence of other sources of incomes of the farmers, from other employment, assets and transfers (pensions, etc.). They may, however, also indicate that farmers working part-time in agriculture simply have lower overall incomes, e.g. because of inadequate farm size and seasonal unemployment. This is shown in the data of Table 11.3 which highlight the large difference in the existence of other gainful employment of part-time farmers in selected countries.

In the United States only 38 percent of aggregate personal income of the farm population originated in farming in 1982 compared with 53 percent in

Table 11.2 Comparative employment and income data in agriculture, EC countries

| Country | Persons with main employment in agriculture, forestry and fisheries * 1987 | Working on agricultural holdings, 1985 | | Net value added at factor cost ‡, average for 1982–6 | |
		Persons	Equivalent full-time workers †	Per person in col.1	Per equivalent full-time worker
	thousand			% (EC-11 = 100)	
Belgium	100	158	107	222	182
Denmark	172	158	122	128	156
France	1489	2246	1565	130	105
Germany, FR	1327	1740	918	68	84
Ireland	164	428	276	120	62
Luxembourg	6	10	7	148	115
Netherlands	245	295	234	252	207
UK	592	713	543	141	144
Greece (1980)	1016	1841	797	88	80
Italy	2169	5134	2126	95	91
Portugal	926	1898	1156		
Spain	1723			73	90
EC-11 (excl. Portugal)	100	100			

Source: CEC (1989a): 48, 49, T95-T100.
* Employment in agriculture as defined in the general employment statistics (used in Tables 2.1 and 11.1).
† Annual work units (AWU).
‡ In purchasing power standards at current prices.

Table 11.3 Farmers working in agriculture less than 50 percent of an equivalent full-time worker, 1985

| | % of total farmers | of which % with other gainful employment | | |
		main	secondary	none
Italy	70	32	3	65
Netherlands	11	60	5	35
Germany, FR	47	73	2	25

Source: CEC (1989a): T102.

1975 (USDA 1983). In the UK, using tax returns data, self-employment income from agriculture and horticulture was 58 percent of total income in 1983/4. Contrary to the picture emerging from the US data (see below), non-farm income provided a higher percentage of total income (68 percent) in the highest income bracket than in all other income groups. In Sweden, again judged on the basis of tax files data, farmers' income from employment (mostly outside agriculture) in 1985 was almost double their income (net receipts) from farming. Non-farm income is also a significant, and increasingly important, contributor to total income in the FRG. Gross income of farm households from the farm business (akin to the concept of net farm income) was 66 percent of total gross income from all sources in 1972 and 52 percent in 1982.

A further issue in evaluating the income position of farmers in the western countries is how to take account of changes in the value of farm assets. In most countries of the region farmers who owned land experienced substantial increases in land values during the 1970s, although land prices have since fallen in response to the reduction in inflation and the downturn in agricultural profitability. In the United States between 1973 and 1983 the addition of capital gains to income would have increased the level of farm income by about 50 percent (Knutson *et al.* 1983). Capital gains and current farm income are not, of course, equivalent; one either has to borrow against capital gains or sell the assets to realize their benefits.

In a policy context the increasing role of non-farm income would tend to mitigate the negative effects of output price reductions on total incomes of the farm population. Further, off-farm income is generally of greater importance to smaller farms, and declines as a proportion of farm income as the size of farm increases. Data for the USA show that when total income of farm families in different size groups is compared to the median income of the total population, the total income of both the smaller and larger size groups exceeds the national median, while only the medium-size categories fall slightly below. These mid-size farms are neither totally reliant on off-farm income nor large enough to achieve comparable farm incomes.

However, even in countries where average farm household incomes are comparable to those in other sectors, pockets of unacceptably low family incomes remain. Some farmers have low family incomes, not necessarily because they are farmers, but because they possess a less valuable bundle of skills and capacities (due in some cases to low levels of education, the poor quality of rural schooling and old age) than other people. The jobs they would obtain in the non-farm sector would also be relatively lowly-paid. Problem farm incomes occur on farms in poorer, remote or mountainous areas which also suffer from additional problems of a structural character such as small farm size. Individual farmers may also suffer income difficulties due to flood damage, disease problems or other disasters, or because of the difficulties of servicing large amounts of indebtedness if interest rates and/or farm profitability vary more adversely (this problem was of particular concern in North America and some Western European countries in the 1980s and is discussed further below). In future, income support policies should become increasingly selective, and a greater effort made to target those individuals and groups most deserving of support in terms of their need, contribution to maintaining the social structure of marginal regions or provision of unpriced environmental and amenity services.

Options to improve the income position of farmers

Price and market policies

The need to reduce policy incentives for increased agricultural production in the western countries in the study (discussed in Ch. 5) implies that price policy will play a less important role in farm income support in these countries in the future. Even in the absence of pressures to deal with the problem of agricultural over-capacity, there is increasing evidence that price policy is an inefficient mechanism of income support with sometimes perverse side-effects. Its across-the-board nature makes it a particularly inappropriate mechanism when income problems are specific to particular groups of farmers.

Price support benefits farmers in relation to the size of their marketed production, and thus large farmers benefit more than small farmers. Whether short-term relative income disparities in the sector are increased or not depends on factors such as the commodity composition of output by farm size and differences in support levels between commodities, on differences in the size of the income multiplier (the ratio between sales and net farm income) across farm sizes and on differences in the adjustment possibilities across farms to the new relative prices. In the longer term, price policy effects on income distribution within farming depend on its impact on the rate and nature of structural change. Some argue that price policy has so strengthened the position of larger farms in the land market that they are able to 'cannibalize' their smaller neighbours and thus hasten the demise of smaller farms.

The income benefits of price support policies may also be transitory, and tend to dissipate through increased prices for certain farm inputs, particularly land. When price support is increased, more people try to stay on or to enter farming because of its greater profitability. As a result, land prices and rents are bid up. The owner-occupier benefits immediately not only from the extra income but also from the capital appreciation of his land which puts him in a stronger position to finance an investment or expansion programme. Farmers who must purchase land to begin or expand in farming are not made better off by higher price supports which raise the cost of land. In countries where rents adjust sluggishly to market conditions tenants are also better off in income terms in the short run. If rents adjust quickly, then the benefits of price support do not remain with the farm operator but are immediately passed on to the landlord.

In addition, in those country groups where agriculture has become more closely integrated with the rest of the economy, e.g. North America and many Western European countries, the effectiveness of price policy in steering farm incomes, even if it were possible to pursue it with a free hand, has become increasingly attenuated as trends on input markets become increasingly important determinants of farm incomes. Above all, the ever expanding role of off-farm incomes of a growing proportion of the farm population will continue to erode the effectiveness of price policies as a means to sustain farmers' incomes.

Price policies may appear to have a more active role to play in income support in CPE countries where increased agricultural production remains an important goal. However, the role of price policy cannot be considered in isolation from changes in the subsidy system which are currently taking place, and the overall effect on net transfers between the agricultural and non-agricultural sectors is not always clear.

Decoupling income support from production

The main alternative to price policy, which may offer a less distorting and more cost-effective means of meeting the objective of income support, is to adopt policies aimed at decoupling income support from production, e.g. by using direct income payments to farmers. Such payments would not depend on the level of production, although they would affect it, in the sense that any payments scheme in which entitlement is tied in some way to farming, will attract, or at least keep, resources in agriculture.

Study countries are gradually accumulating a body of experience with this form of income transfer. Initially, payments were targeted to particular groups of farmers such as those in disadvantaged or mountain areas. Increasingly, direct income supports are being used to compensate for reduced levels of price guarantees. The USA moved to a system of deficiency payments for grains and cotton which, when combined with output and payments limitations (although recently relaxed), can be considered in this context a move towards decoupling income support from production. Governments have also made income support payments in response to crisis conditions brought on by poor weather, a steep rise in production costs, etc. In addition, in both CPE and market economies net transfers are made to agriculture through social and fiscal channels which, while not explicitly granted as income support, none the less contribute to improving the income position of farmers. A more novel approach to direct income support is to pay farmers for their contribution to amenity, recreation, the availability of an attractive landscape and wildlife habitat. Direct payments to farmers for environmental conservation activities are a means of encouraging them to take the public interest in the maintenance of these services into account.

These different objectives of direct payments — payments in recognition of social need, compensation for losses imposed by changes in government policies, the alleviation of hardship or payment for unpriced environmental services — are reflected in the different bases proposed to gain eligibility for support. If the primary concern is social equity, then ideally payments should be differentiated in accordance with the income level and possibilities of the farm household, although the administrative requirements to permit differentiation may be overwhelming. A negative income tax scheme could perhaps meet these requirements, but in only a few countries are farmers fully integrated into the income tax system. It must be recognized, however, that no scheme will be free of unwanted side-effects which, with the passage of time, may overwhelm and compromise the achievement of the primary objective. Further, once the basis of support shifts from land or products to people, comparisons with the situation of those in the non-agricultural sector become inevitable. It may not seem fair to treat farmers more generously than other workers displaced by technological change. If income payments are designed to provide transitory compensation to farmers for price reductions brought about in response to the budget implications of agricultural policy, they have the disadvantage that the peak demands for income support occur from the outset whilst savings to the budget from lower prices take a long time to materialize fully; more budget resources, not less, are required in the short run. The US Treasury experience with deficiency payments for grains and cotton in the 1980s is a good example of this. Expenditure can be limited by setting a ceiling on the maximum payment per farmer, an approach which has much to commend it on equity grounds but which may draw substantial opposition from the larger farmers who are adversely affected.

Part-time farming

An alternative approach to maintaining farm incomes is to encourage pluriactivity among farmers. In countries such as the United States, Austria, Norway and Switzerland more than half of all farms are part-time farms. A recent study indicates that this is also the situation for nearly half of the farms in the UK (MAFF 1988). Table 11.4 indicates the distribution of holdings in the EC by full-time, part-time (defined here as between 50 and 100 percent of holder's working time worked in the farm) and spare-time (less than 50 percent) categories. Assuming that how the holder spends his working time is a good indicator of how the farm is managed, the majority of holdings in the Community are less than full-time operations and in Italy and Greece most holdings are spare-time. The term 'spare-time' is probably not appropriate to the situation in these two latter countries: the high shares more likely reflect the smallness of farm size which absorbs only a minor proportion of available working time, not of time effectively devoted to gainful employment. Also definitional differences among countries play a role: many of the small farms in Italy and Greece would in the other countries probably not be counted as farm units. From an incomes policy viewpoint, the importance of part-time farming lies partly in the fact that it tends to be concentrated, although by no means confined to, smaller farms. It thus tends to have an equalizing effect on the distribution of the total household income of farm families as well as providing a means of lifting income of many farm households above minimum levels.

Government attitudes to part-time farming are determined by whether this phenomenon with its many ramifications outside of agriculture fits in or not with their broad agricultural policy goals. When part-time farming seems to be consistent with achieving certain broader policy goals, e.g. those concerning demographic spread, regional development and preservation of the rural environment, government atttitudes tend to be favourable. Where governments place primary emphasis on increasing production and

Table 11.4 Holders by proportion of normal working time on the farm, 1983 (percent of working time worked on the farm)

	Below 50 (spare-time)	More than 50, less than 100 (part-time)	100 (full-time)	Number of holders (000s)
Germany, FR	47.2	8.4	44.4	750
France	29.0	15.5	55.5	1,075
Italy	72.5	15.9	11.6	2,760
Netherlands	11.3	13.8	74.9	135
Belgium	28.3	7.1	64.6	100
Luxembourg	18.6	9.3	72.1	4
UK	25.0	11.9	63.1	220
Ireland	27.2	26.9	45.9	214
Denmark	21.7	16.4	61.9	96
Greece	65.7	24.1	10.2	957
EC-10	55.1	16.2	28.7	6,313

Source: Eurostat. Quoted in Robson *et al*. 1987.

structural reform towards a system of viable farm holdings is important, part-time farming is often viewed in a less favourable light. It was already noted in Chapter 8 that pluriactivity is widespread in the CPEs and is gaining in importance, e.g. in the form of household plots and auxiliary farms which provide additional income not only to the employees of the state farms and kolkhozes but also to persons working in the non-agricultural sector.

The further extension of dual jobholding will depend on increased employment opportunities, both on-farm and off-farm, in rural areas. Government rural and regional development programmes can assist this process. For example, new on-farm activities such as the provision of accommodation services or food processing activities or so-called 'niche' productions might be encouraged. Off-farm employment opportunities can be created through incentives for industrial relocation and the improvement of public infrastructure. Participation in non-agricultural job opportunities is selective, however, with elderly farmers and farm families with lower education levels at a disadvantage, so part-time farming will remain at best only a partial solution to the low-income farm problem.

Price and income instability

In addition to the income parity objective, many governments have intervened to even out price and income fluctuations over time and to bring about greater price and income stability, an objective of high priority in nearly all countries. In the western countries in the study the variability of both gross and net farm incomes increased during the 1970s (OECD 1981). Whereas policy in the past emphasized domestic sources of instability, and particularly domestic supply shifts due to weather and similar factors, in the

1970s product market instability increased mainly because of the additional instability introduced by international trade and exchange rate developments. In addition, input cost instability, particularly in credit, energy and land markets, made a much more important contribution to income instability than previously.

The farm credit crisis in North America and some Western European countries in the 1980s is a case in point. Many farms increased their level of borrowing significantly in the 1970s, encouraged by negative real interest rates and by forecasts of tight supply–demand balances for agricultural products in the future. Higher land prices also resulted in higher levels of debt for farmers who bought land. These highly leveraged farms were very vulnerable to the high real interest rates and falling asset values at the beginning of the 1980s (in the USA, the ratio of debt to total assets in the farm sector averaged about 16 percent in the 1970s; by 1985 it had reached 24 percent (USDA 1986a). The general farm-price/input-cost squeeze also reduced the cushion available to farmers to meet higher debt servicing costs. Many governments introduced emergency interest rate reduction packages in the early 1980s in response to the cash flow crisis on these farms.

This example illustrates that the stabilization of product prices is now a much weaker tool of income stabilization, and may be further attenuated by the results of the round of agricultural trade negotiations now under way. The search for alternative methods of income stabilization will be an important preoccupation of governments in the coming period. Farmers themselves make use of a number of arrangements to cope with market instability. These include enterprise diversification, prudent financial management, the use of provisions in the taxation code which permit income averaging, and forward contracting and futures markets. Alternative mechanisms include the use of insurance programmes to protect crop producers against natural hazards, greater provision of market intelligence services for producers and minimum price guarantees at levels which protect producers against serious losses without permanently raising producer prices.

Conclusions

Price and market guarantees have been the principal measures used to support and stabilize farm incomes, particularly in the western countries in the study, although the relationship between price policy and standards of living in the farm sector is not often evaluated explicitly in policy formulation. The need to reduce these price guarantees in the context of continuing overcapacity in the agricultural sectors of these countries means that governments will turn increasingly to other mechanisms of income support. Statistics on farm household income (including income from off-farm employment) and non-farm household income suggest that, at least for the CPE countries and the higher-income market economies, rough parity has been established at the average level and that remaining farm income problems are concentrated in particular farming groups and regions. Further price reductions, particularly if they are substantial and take place over a short period of time, will cause additional transitional difficulties for farm incomes, although also here their effect will be mitigated for many farms by the proportion of household income coming from off-farm sources. Thus future income support policies should be increasingly selective, and should make a greater effort to target those groups and individuals most deserving of support because of their need or contribution to other agricultural policy objectives.

It is clear that direct payments in various forms as well as off-farm income will make up an increasing proportion of farm household income in the coming years. The administrative burden of direct income support schemes should not be underestimated, and it is likely that an increasing variety of approaches will be tried. It is often said that farmers are unwilling to accept direct payments, on the grounds that they represent a form of hand-out and are thus demeaning. Experience with income support schemes to date (though admittedly many still tie eligibility to some form of production) suggests that this problem has been overstated in the past. A bigger difficulty might be the budget cost of transferring income support from consumers to taxpayers. Although ceilings can be put on the size of payments to individual farmers to reduce the budget exposure, such limitations may lead to a loss of political support in farmer organizations.

12 Improving the performance of the agricultural trading system

Trade policy objectives

The central theme of this study has been an examination of policy options to bring about agricultural adjustment. Advances in agricultural technology, the pace of which some predict may quicken in the next two decades (see Ch. 7), imply a continuing increase in agricultural resource productivity. But the markets for absorbing rapidly growing production do not exist, as final consumption is expected to grow more slowly and export markets (to countries outside the region) will be less buoyant than in the past. Thus, in most market economies of the study, but also in some CPEs, the resources employed in agriculture need to grow more slowly, or to shrink more rapidly, than in the past. In addition, some countries within the region intend to increase their levels of food self-sufficiency (some CPE countries) and/or their market share by exploiting their comparative advantage in agricultural production (Southern Europe) where this is possible. This will further intensify the pressures for resource adjustment and reduction in other countries in the region.

Market imbalances have been exacerbated by policy interventions in many of the countries studied which, although often introduced in pursuit of non-production objectives, have had production and demand implications. The subsequent disequilibria have been reflected in high budget and consumer costs, and in the widespread disarray in world agricultural trade.

World agricultural markets have always been sensitive to small fluctuations in global supply–demand balances, but the restrictions on access to actual or potential deficit markets and the large surpluses of exporting countries have resulted in widespread distortions in trade patterns and in price levels. In the negotiating objective for agriculture established in September 1986 for the Uruguay Round of Multilateral Trade Negotiations (MTNs) the GATT Contracting Parties agreed that there was an urgent need to bring more discipline and predictability to world agricultural trade by correcting and preventing restrictions and distortions, including those related to structural surpluses, so as to reduce the uncertainty, imbalances and instability in world agricultural markets. They further agreed that 'negotiations shall aim to achieve greater liberalization of trade in agriculture and bring all measures affecting import access and export competition under strengthened and more operationally effective GATT rules and disciplines' (see below for further developments and the agreement on agriculture reached at Geneva in April 1989 during the Mid-Term Review of the Uruguay Round).

The particular problems of liberalizing agricultural trade are well–known.

In many of the countries of this study agricultural trade policy exists fundamentally to complement domestic support policies designed to redistribute income to agriculture and to contribute to other domestic objectives. Thus the reduction in agricultural trade barriers requires the adjustment in varying degrees of these domestic objectives unless equally effective, non-trade distorting measures are put in place. Also, protective measures rely heavily on the use of non-tariff barriers (NTBs, e.g. quotas, voluntary export restraints, variable import levies) which make the degree of support much less transparent and make agreement on a package of mutually agreed concessions more difficult to reach. Their discrete nature also makes their gradual removal (as can be done with tariffs) more difficult. Options considered in the process of policy reform towards trade liberalization include proposals to convert such NTBs into tariff equivalents (tariffication) whose gradual reduction can be negotiated.

The need for agricultural policy reform agreed at Punta del Este as well as at the OECD Ministerial Council Meeting in May 1987 and at the Mid-Term Review of April 1989 (see below), has become more urgent. Some actions to correct market imbalances have already been undertaken. Friction in agricultural trade continues high and threatens to extend to other products, sparking a full-scale trade war. The study countries account for a high proportion of world agricultural trade, as well as a high proportion of the global amount of assistance given to agriculture, so the agricultural trade negotiations in the Uruguay Round take on a particular significance for them.

Trade policy options

While all parties to the GATT have accepted the need for both greater liberalization and more discipline and predictability in world agricultural trade, there remain wide differences between countries in how best to achieve these objectives. These differences concern the objectives, extent and modalities of policy reforms, the phasing of such reform and attitudes to the strengthening of GATT rules and procedures concerning agricultural trade. The fundamental difficulty is how to address the problem of overcapacity in the agricultural sector, whether by lowering agricultural support or by taking steps to blunt its consequences for world markets.

The central questions in the GATT negotiations relate to the ultimate extent of liberalization and the means for its achievement. The principle of the phasing-in of policy reforms is generally accepted (progressive reduction in agricultural support and protection), although concrete details have yet to be worked out, e.g. the timing and pace of implementation, and how to account for policy changes undertaken to date. However, discordant views as to the ultimate objective have continued to prevail, i.e. complete phasing-out of all trade-distorting support and protection within an agreed time horizon or progressive reduction, even if substantial. Such discordant views were at the root of the failure to reach consensus in the area of agriculture at the mid-term ministerial meeting on the MTNs in Montreal in December 1988. This resulted in putting 'on hold' all agreements on future MTN work pending reaching a consensus on how to proceed in the area of agriculture (as well as in textiles and clothing, intellectual property rights and safeguards). Such consensus was reached in April 1989 (for a description and analysis see Hine *et al.* 1989). It is a compromise which makes no reference to time-bound complete elimination of agricultural support and protection with trade-distorting effects. Instead, the agreed definition of the long-term objective of policy reform is to 'establish a fair and market-oriented agricultural trading system' and 'to provide for substantial progressive reductions in agricultural support and protection sustained over an agreed period of time, resulting in correcting and preventing restrictions and distortions in world agricultural markets'. The United States and the Cairns Group[1] have proposed the removal or virtual removal of all subsidies and import barriers which distort agricultural production and trade. Thus, in the long run, the only measures of agricultural support which would be compatible with these proposals are those measures which are neutral with regard to the impact on the level of domestic agricultural production and which do not distort trade.

Other countries and groups of countries, including Japan, Switzerland, Austria, the EC and Nordic Countries tend to regard trade imbalances as the basic problem, and the remedies which they propose do not envisage complete liberalization of agricultural markets nor total abolition of market support and protection. Their proposals differ in a number of aspects, though in practical terms they envisage action towards correcting market imbalances through reduction in domestic support and protection (S & P) and border measures, including export subsidies. While these countries generally envisage some scope for greater use of tariffication and decoupled support, a common feature of their proposals is an emphasis on 'non-economic' aspects of domestic policy goals such as food security, environmental and conservational, and regional concerns, to be achieved also by price policy measures. In the specific case of the EC proposal it is envisaged that correction of market imbalances should also entail a 'rebalancing' of support and protection, to be accomplished by raising S & P on certain products within the context of an overall reduction of agricultural support.

It is probably true that if all S & P were removed, some European countries would experience declines in the production of some commodities, eventually turning from net exporters to net importers. Most studies on the effects of removal of S & P do indeed predict such declines although at greatly varying rates. For instance, Harvey and Hall (1989) estimated the production changes that would have resulted in 1986 if the agricultural support system had been operated on the basis of production entitlement guarantees (PEGs) fixed at both the national and the farm level (see below for a discussion of the PEG system). According to the authors, under this 'quasi-non-distorting policy alternative....the incentive price on both the supply and demand sides of the market will be the free-trade market price' with support payments limited to supply quantities which are

---

[1]   Argentina, Australia, Brazil, Canada, Chile, Colombia, Fiji, Hungary, Indonesia, Malaysia, New Zealand, Philippines, Thailand, Uruguay.

no greater than free-trade ones. For a better appreciation of these potential effects, see Table 12.1 which shows the estimated percentage changes from 1986 levels together with the actual production, net trade and self-sufficiency data for the three-year annual average 1984–6. The greatest declines would have occurred in sugar, beef and veal and milk, with the EC-10 turning from net exporter to net importer. Smaller declines are predicted for wheat and barley, with the Community remaining a net exporter, and increases for pigmeat and eggs with the EC increasing net exports.

Table 12.1 Simulated changes in 1986 production and actual levels for 1984–6, EC-10

| | Changes in 1986 production under PEG support policies,% | Actual levels, annual average 1984–6 | | |
| | | Production | Net trade | SSR [†] |
		million tons		%
Wheat	−11.0 [*]	69.6	14.0	134.3
Barley	−3.5	41.5	8.8	129.2
Sugar	−28.7	13.6	2.9	129.4
Pigmeat	17.7	10.6	0.3	102.4
Poultry	−6.6	4.4	0.4	107.1
Eggs	6.1	4.2	0.1	102.2
Beef and Veal	−23.3	7.5	0.5	108.7
Sheepmeat	−9.3	0.7 [‡]	−0.2 [‡]	77.1 [‡]
Milk	−23.4	117.0	11.0 [§]	110.9 [§]

Sources: Column 1: Harvey and Hall (1989); other columns: FAO.
*   Common wheat only (about 90 percent of all wheat production).
†   Production percent of domestic use.
‡   Sheep and goat meat.
§   Fresh whole milk equivalent of all dairy products.

Another recent study by Roningen and Dixit (1989) predicts that under multilateral liberalization in the industrial countries (complete elimination of S & P levels prevailing in 1986/7) producer incentive prices (prices received plus direct support payments) for wheat would fall by 44 percent in the EC and by 35 percent in the rest of Western Europe. As a result, production would decline by 16 percent and 15 percent respectively. For coarse grains the declines are 34 percent and 37 percent for prices and 4 percent and 10 percent for production, respectively. These effects of the removal of S & P reflect the results of static analysis and represent the new equilibrium that would be reached after about five years of adjustment 'with the important proviso that all other conditions remain the same as in the base year 1986/7' (p.11).

This *ceteris paribus* clause may not hold when the situation is examined in a dynamic context. It was noted earlier (Ch. 5) that producer price declines did occur over time and that they have failed to arrest the growth of production, due mainly to the strong productivity trends. If an allowance is made for such trends, the production declines need not occur in many commodity/country situations to the extent predicted by the models. A reduction in S & P may be partly absorbed in the form of lower returns to land (land values and rents) and it may also intensify the process towards the emergence of larger and more efficient units, leading to further increases in the already high share of such units in total output, a process that has been taking place under the extended regime of S & P (see Ch. 8). In this process, small marginal farms will probably suffer, and this is one of the reasons for the reluctance of many countries to contemplate elimination of S & P. But in many cases the fears that European producers will find themselves at a competitive disadvantage *vis-à-vis* those of the countries with more abundant land resources may not be wholly justified. Stanton (1986), who carried out a comparative analysis of cereals production costs in the major EC countries and the USA for the early 1980s concludes that, with all the caveats applying to such comparisons,

the capacity of EC producers to compete in world export markets for wheat with North America, Australia and Argentina on technical grounds is substantial. High internal prices have encouraged this major growth in cereal production and modernization of plant and equipment. Now that it is in place, it is a competitive force in terms of production costs in world markets. Public policy-makers and wheat producers in North America must come to recognize that EC producers will continue to generate an export surplus even if internal prices continue to fall toward world market levels. (p.172)

The earlier chapters of this study reviewed the policy options for agricultural policy reform, including for their effects on production, consumption and trade, which are of particular interest to the topic of this chapter. Some recent contributions have focused more closely on the analysis of policy options from the standpoint of their compatibility with the objectives of increased market orientation and trade liberalization (FAO/CCP 1989; Gardner 1988; IATRC 1988; Blandford *et al*. 1989). Among these are policy options involving direct payments to farmers per unit of output unrelated to current production, e.g. when the farmer is entitled to receive a fixed or a deficiency amount per unit of his/her average production in some historical period. Alternatively, such payments may be made to each farmer on only part of his/her current output (production entitlement guarantee — PEG) with the proviso that the sum-total of production thus benefiting would be less than aggregate 'free market' output, i.e. output at prices which will no longer be supported through intervention or border measures. For this PEG option to be considered as not influencing production decisions, payments must be guaranteed irrespective of whether the farmer produces his entitlement or not. These schemes, of course, imply that all farmers (and non-farmer land owners) must be compensated for the reduction or elimination of the S & P policies which distort markets. While this may be politically expedient at times, the schemes provide the flexibility of discriminating among farmers on the basis of criteria of need or location. They would also make it possible for governments to reduce the aggregate guaranteed production, or to better target income distribution and regional development objectives, by offering to buy up and extinguish or reassign part of the outstanding PEGs (for discussion of the capitalized value of entitlements to receive support see Gardner 1988).

Even before the initiation of the Uruguay Round considerable attention had been given in the GATT to ways in which agriculture might be brought into the framework of rules and disciplines of the General Agreement, notably by the Committee on Agriculture established to pursue the work programme agreed at Ministerial level in 1982. In the Uruguay Round, complementary activities of the Negotiating Group on Agriculture have included a focus on ways of developing a quantitative, 'aggregate measurement of support' provided to agriculture, for use as a negotiating instrument or for monitoring eventual commitments made on policy reform. The conceptual basis of such a measurement is provided by the Producer Subsidy Equivalent (PSE), developed in FAO, the GATT and the OECD (for discussion see FAO 1973, OECD 1987 and Tangermann *et al.* 1987). Widespread but not universal agreement has been reached on the need for some quantification along such lines. However, agreement has yet to be reached on its common definition, including such aspects as commodity and country coverage, and the elements of support policies which should be included in or excluded from the measurement, and on the possible scope of its application in the negotiations.

The choice of strategy by individual countries in the trade negotiations will reflect the different weights attached to the various goals of agricultural policy. Given that agricultural trade policy is often the external face of domestic support policies, it is hardly surprising that preferred strategies for domestic policy reform will be reflected in the approach to trade negotiations. But the attempt to seek a coordinated approach to trade problems has very real benefits for participating countries. First, it eliminates the fear of countries contemplating reforms that other countries will take advantage of their unilateral efforts. Second, the adjustments required by any one country are less if all countries simultaneously reduce their support to agriculture. For example, under multilateral liberalization, world prices are likely to rise and, therefore, national measures to compensate farmers for income losses, if any, will be less costly than under unilateral reduction in support and protection. It is thus very appropriate that the Uruguay Round should be the focus for efforts to reform agricultural policies.

Implications of policy changes for developing countries

The agricultural and trade policies of countries in the study region affect developing exporters primarily by restricting their market access. An effective curb on the demand for their exports to the CPEs is the limited availability of hard currencies. In addition, state monopolies in foreign trade insulate the domestic market from competition from foreign sources, in keeping with the drive to greater self-sufficiency, and can prevent potential consumer demand for tropical products finding expression. As a result, the CPEs account for only a small share of agricultural exports from the developing countries, except for those with special relations with the CMEA. In the rest of the study countries, access to markets in which certain developing countries have a comparative advantage (in particular, sugar, meats, fruits, vegetables, oils and some feedstuffs) is restricted by a variety of tariffs, levies, non-tariff barriers and, mostly for the non-competing products, domestic taxes. Another concern of the developing countries is the volatility of world commodity markets. The vagaries of nature on the supply side and cyclical variations in demand are important sources of instability. However, interventionist policies aimed at insulating domestic markets also tend to make for greater instability in residual world markets.

Because of the interdependencies between agricultural products in both production and consumption, the direction of impact of domestic policies of the industrial countries on the world price levels for individual commodities is not always immediately obvious, though most studies on the subject indicate that the net effect has been that prices in world markets are lower and more volatile than they would have been otherwise. By implication, exporting developing countries have suffered a loss by exporting less and at lower prices, while there is a prima-facie case that importing countries have gained from the more favourable terms of trade. In practice, many developing countries are both exporters (e.g. of sugar) and importers (e.g. of cereals) of the commodities whose world market prices have been adversely affected by S & P policies in the industrial countries. Additionally, a number of preferential access arrangements by some developing countries to the protected markets of some industrial countries further complicates the picture (e.g. the ACP countries' access to the EC market under the Lomé Convention, or the higher than world market prices for sugar exports to the USA under the import quota regime — a dubious and rapidly disappearing privilege however, see Maskus 1989). These factors make identification of gainers and losers on the basis of the simple importer–exporter criterion difficult in many cases, though there exist straightforward cases of losers (e.g. the actual or potential exporters of cereals and livestock products or the exporters of tropical competing products, particularly exporters excluded from preferential access arrangements) and gainers (e.g. mineral exporters who are also large importers of temperate zone foods or countries who benefited from plentiful cereals supplies through concessional sales and food aid).

Further, when general equilibrium aspects of trade distortions are taken into account, low world market prices are believed to be a factor making for unfavourable terms of trade of agriculture in the developing countries leading to decreased profitability for the sector and adversely affecting development priorities and food self-sufficiency. Some analyses also show that the resource-use inefficiencies inherent in the agricultural S & P regimes in many industrial countries result in high costs for the non-agricultural sector and, therefore, in higher prices of manufactures exported to the developing countries. By the same token, some developing exporters of manufactures (e.g. the NICs) must have benefited from this situation.

Much of the preceding discussion derives from analyses of the possible production and world market price effects of trade liberalization scenarios in the industrial countries. Particular attention has been paid to the impact on the world market prices of cereals and their implications for the food security of the food deficit developing countries. A survey of these issues is given in FAO/CFS (1987), while the discussion is further developed in

Konandreas and Perkins (1989). Much of the following discussion draws on this latter paper. According to the authors, the most commonly accepted conclusion of the trade liberalization studies is that, at least in the short-term, cereal prices will rise. Such a rise could be substantial if liberalization is confined to the OECD countries only (the above mentioned study by Roningen and Dixit predicts an increase in world market price of 37 percent for wheat, 26 percent for coarse grains and 65 percent for dairy products). It would be more modest if also the developing countries liberalized on the premises that at present they have negative protection for their agricultures, both as a result of sector-specific policies as well as economy-wide ones (on this latter point see Krueger *et al*. 1988). Indeed, one of the latest studies on the subject concludes that complete elimination of (positive and negative, direct and indirect) S & P to agriculture in both the industrial and the developing countries would result in a decline in world cereal prices because of the greatly increased incentives to the agriculture of the developing countries (Zietz and Valdés 1989).

Whether world cereal prices rise substantially, modestly or even remain constant, liberalization would remove structural surpluses in the major exporting countries and this can be expected to reduce concessional exports to the developing countries, both food aid and variously subsidized commercial sales. This would create serious difficulties for the many low-income food-deficit countries which depend substantially on such concessional imports for their food supplies. Some of these countries would be compensated through the higher prices for their agricultural exports if their prices also rise as a result of trade liberalization. Others may not be so lucky and those who cannot substitute domestic production for the more expensive imports (because of ecological and other reasons) would be hit the worst. In all cases the severe foreign exchange scarcities faced by many of these countries would make it difficult to allocate increased amounts to maintain food imports at the previous levels. Additionally, there would be pressures to compensate at least low income consumers who are net buyers of food for the higher food prices, something that the overburdened budgets of these countries can ill afford under the present policies for structural adjustment. It is conceivable that some revenue for this purpose can be generated by taxing some of producer gains resulting from the higher market prices, but the scope for this may be limited (in the typical African low income food-deficit country only about a quarter of domestic cereal output is marketed — the balance being subsistence, therefore the bulk of producers may not benefit from the higher market prices). Such a course would also negate the incentive effect of higher prices on domestic production.

Faced with this problem, the policy responses at the international level may take the form of (a) ensuring that trade liberalization benefits to the maximum extent possible and also the commodities exported by these countries, and (b) some form of compensation to the particularly needy countries to enable them to continue to import at nearly the previous levels. The scope for the former option is substantial for some products and countries (e.g. sugar, beef, fruit) but rather limited for countries exporting mainly non-competing tropical products (see Box 12.1). The latter option should be made easier by the realization that liberalization in the industrial countries would increase their overall economic efficiency, producing gains of which a small part could finance such compensation transfers. In practice, the existing S & P policies do indiscriminately subsidize foreign importers, rich and needy alike, through subsidized exports. Only part of the savings from policy reform may be required to compensate needy importers for a transitional period.

Box 12.1: *Trade policies and developing countries: the case of tropical products*

In the Uruguay Round of Multilateral Trade Negotiations, as in the two previous GATT rounds, tropical products are to be afforded priority. The bulk of exports of most tropical products originate in the developing countries which, therefore, stand to benefit from early liberalization. Most tropical products compete with production in the major developed countries, either directly e.g. vegetable oils, or indirectly, e.g. Third Country bananas with EC/ACP fruit, manioc with cereals, natural rubber and jute with synthetics. In this connection, the recent EC proposal to the GATT on trade liberalization covered a limited number of tropical products (tropical beverages, tropical plants, tobacco, tropical fruits and nuts, tropical wood and rubber, jute and hard fibres) and excluded some key products which compete with domestic production.

The distinction between *competing* and *non-competing* products is important. The scope for increased exports from the low-cost developing country exporters is much greater for the former than for the latter class of commodities: exports of competing products would increase mainly by displacing high cost production in the importing countries following trade liberalization. For the non-competing products, however, the developing countries are the main or sole producers and, therefore, the volume of their exports would expand only to the extent that liberalization would stimulate consumption. Additionally, they may gain from price increases and from exporting a higher proportion in processed form.

For the non-competing products gains in export earnings from *price and volume effects* are likely to be modest, at least for those exporting the raw product. A study (Valdés and Zietz 1980) estimated that in 1975–7 an across-the-board 50 percent reduction in protection in 17 developed countries (of which 14 countries are in the European Study region) would have increased earnings by only 3.1 percent for green coffee exports, 5.0 percent for those of tea and 2.1 percent for cocoa beans exports, as major markets of Western Europe and North America are relatively saturated. However, if trade liberalization was sufficiently widespread and particularly if the CPEs were included, the impact on the volume of trade would be greater, although tempered by the rise in world market price which such a shift in world demand would induce.

The scope for gains in export earnings from non-competing products by shifting to production of *processed goods*, following trade liberalization, will depend very much on the type of product under review. Specifically, as the value added in processing tropical beverages is relatively small, the potential gains from increased domestic processing are lessened. There are a number of studies that have attempted to estimate gains in export earnings from exporting processed rather than primary tropical beverages (for examples see MacBean and Nguyen 1987). A recent study by Mabbs-Zeno and Krissoff (1989) estimates that if the developed countries had eliminated in 1986 their policies which restrained consumption and, through escalating import barriers, promoted domestic processing, export earnings of the developing exporters would have been higher by $348 million (or 21 percent) for processed products (coffee and cocoa) while earnings from primary exports of these two products plus tea would have changed very little. The net result would have been a net increase in earnings from both primary and processed exports of these three products of only 2.4 percent. If, by contrast, the developing country exporters of these products 'liberalized' their policies (essentially removing existing tax 'wedges' between export and producer prices) they would collectively suffer a catastrophic decline in their export earnings—by $4.3 billion or 27 percent of their 1986 total. Some exporters may gain, but the essence is that the higher prices received initially by the producers would lead to increased production and, in the face of low price elasticities of demand, would drive prices down and reduce total export receipts. These results, coming from a modelling exercise similar to that indicated earlier (Roningen and Dixit 1989), should be interpreted with caution but they give food for thought.

Hence, if trade liberalization in the industrial countries is confined to relatively non-competing primary tropical products, the LDCs will undoubtedly benefit but their gains cannot be expected to be large. Large trading gains will be found in the liberalization of trade in those competing products and value-added processed products subject to escalating trade barriers, in which developing countries have competitive advantages. Moreover, long-term development will require a greater diversification of the developing countries economies and so trade liberalization should be sought over an even wider range of export products, both agricultural and industrial—not just the tropical products currently of interest to the developing countries.

Conclusions

Given that agricultural trade policy is often the external face of domestic support policies, it is hardly surprising that preferred strategies for domestic policy reform will be reflected in the approach to trade negotiations. But the attempt to seek a coordinated approach to trade problems has very real benefits for participating countries. First, it eliminates the fear of countries contemplating reforms that other countries will take advantage of their unilateral efforts. Second, the adjustments required by any one country are less if all countries simultaneously reduce their support to agriculture. It is thus very appropriate that the Uruguay Round should be the focus for efforts to reform agricultural policies, also because the Uruguay Round provides an opportunity for the interests of the developing countries to be taken into account when trade and agricultural policies are being negotiated.

The impact of trade liberalization on the developing countries needs to be studied in more depth and detail than has been done so far. There are unequivocal cases when benefits to a large number of commodity/country situations, from liberalization in the industrial countries and in the developing countries themselves, can be clearly identified. Indeed, how much the developing countries will benefit from liberalization in the industrial countries will also depend to a significant degree on their own action to reform policies aimed at removing disincentives to their own producers. There are also clear cases of damage that some developing countries may suffer, at least in the short to medium term, following liberalization in the industrial countries. The developing countries as a whole may also suffer losses if they themselves relaxed some policies which currently provide price support in international markets for non-competing tropical products.

Appendices

1. Country and commodity classification

2. Methodology and projections: a summary note

3. Statistical tables

Appendix 1: Country and commodity classification

List of countries and country groups

EC-north (1)	EC-south (2)	Scandinavia (3)	Alpine countries (4)	Non-EC Mediterranean (5)	North America (6)	European Centrally Planned Economies (CPEs) (7)
Belgium	Greece	Finland	Austria	Albania	Canada	Bulgaria
Denmark	Italy	Iceland	Switzerland	Cyprus	United States	Czechoslovakia
France	Portugal	Norway		Israel		German DR
Germany, FR	Spain	Sweden		Malta		Hungary
Ireland				Turkey		Poland
Luxembourg				Yugoslavia		Romania
Netherlands						USSR
United Kingdom						

Other country groups:

FAO European Region: all countries in groups (1) to (5) and (7) excluding the German DR and the USSR.
OECD member countries: all countries in groups (1), (2), (3), (4) and (6) plus Australia, Japan, New Zealand and Turkey.
CMEA member countries: all countries in group (7) plus Cuba, Mongolia and Viet Nam.
Developed Market Economies (DMEs): all countries in groups (1) to (6) plus Australia, Japan, New Zealand and the Republic of South Africa.
Western Europe: all countries in groups (1) to (5).
Eastern Europe: all countries in group (7) excluding the USSR.
EC-12: all 12 EC countries.
EC-10: EC-12 excluding Portugal and Spain.
EC-9: EC-10 excluding Greece.

Notes: Country groups (1) to (7) are used for analytical and presentational purposes only.
Data for Belgium and Luxembourg are shown together in most tables.

Commodity coverage and specification

The commodities and commodity groups used in the analysis are shown in the list. They are those of the study *World Agriculture: Toward 2000*. Some of these commodities are not produced in the countries of this study, e.g. tropical beverages, cassava, rubber and tropical oilseeds, but are covered for purposes of analysis of demand and trade. Likewise, data for some important commodities of European and North American regions are used individually in the analysis and shown in the Statistical tables, e.g. oats, rye, rapeseed, cotton seed, olive oil.

The weights used for aggregating the physical quantities of commodities into volume totals (e.g. total agricultural production, total demand) and for computing the corresponding growth rates or self-sufficiency ratios are the prices for 1979/81 used for constructing the FAO production index numbers for groups of countries. Each commodity is assigned the same price in all countries (for their derivation see any recent *FAO Production Yearbook)*.

All commodity data and projections in this report are expressed in the specification given in the list, which is usually the primary-product equivalent, unless stated otherwise. Historical commodity balances (Supply Utilization Accounts—SUAs, see Appendix 2) are available for about 170 primary and 200 processed crop and livestock commodities. The full list is available in FAO, *Towards a World Agriculture Information System,* Rome, 1988. To reduce this amount of information to manageable proportions, all the SUA data were converted to the commodity specification given in the above list, applying appropriate conversion factors (and ignoring joint products to avoid double counting, e.g. wheat flour is converted back into wheat while wheat bran is ignored in the construction of the SUA for wheat; production of cereal brans resulting from domestic food use of cereals is, however, estimated and used in the calculation of cereals feed requirements). In this way one SUA in homogeneous units is derived for each of the commodities of the study. Meat production refers to indigenous meat production, that is production from slaughtered animals plus the meat equivalent of live animal exports minus the meat equivalent of all live animal imports. Cereals demand and trade data include the grain equivalent of beer consumption and trade.

The commodities for which SUAs were constructed are 25 crops (nos. 1–25 in the list) and six livestock products (nos. 26–31). The production analysis was, however, carried out for 30 crops because sugar and vegetable oils are decomposed (for production analysis only) into the crops shown in the footnote to the list.

The existence of joint products and by-products makes it impossible to avoid distortions in the consumption and trade data when a large number of commodities are converted into a smaller set of equivalent (mostly primary) products. For example, in the case of dairy products expressed in fresh whole-milk equivalent, any country importing and consuming skimmed milk powder (SMP) is shown as importing and consuming the amount of fresh whole milk that went into the production of the SMP. Trade and/or consumption (perhaps in another country) of the joint product—butter—is not shown at all to avoid double counting and also because animal fats are not covered in this study. Oilseeds are another commodity with significant joint products (oils, oilmeals) and the associated problems of allocation of consumption and trade. All consumption and trade of oilseeds, oils and the downstream products of oils are recorded in oil equivalent while the joint product oilmeal is ignored. Thus any country importing only oilmeals is not recorded in the main data system as trading in oilseeds, although the relevant analysis is carried out in an ad hoc manner outside the SUA data set.

The above examples indicate that further work on the data and methodology should aim at improving the commodity coverage and conversion system used in this study. It would involve not only an increase in the number of commodities (e.g. addition of oilmeals and animal fats) but also some changes in the model structure to account for the fact that production of some primary products (e.g. soybeans) depend simultaneously on developments in two separate markets, e.g. that for oils and fats and that for feeding stuffs.

List of the commodities

Commodity	Aggregation weights (per tonne)	Commodity aggregates	
		Name	Commodities included
1. Wheat	148	Coarse grains	3–7
2. Rice	207	Cereals	1–7
3. Maize	126	Meat	26–9
4. Barley	120	Livestock products	26–31
5. Millet	129	Non-food products	19–25
6. Sorghum	126	Total food products	1–18,26–31
7. Other cereals	121	Total agriculture	1–31
8. Potatoes	107		
9. Sweet potatoes	85		
10. Cassava	80		
11. Other roots	118		
12. Sugar, raw *	247		
13. Pulses	364		
14. Vegetables	177		
15. Bananas	130		
16. Citrus fruit	167		
17. Other fruit	237		
18. Vegetable oil and oilseeds (vegetable oil equivalent) †	1131		
19. Cocoa beans	1207		
20. Coffee beans	1261		
21. Tea	1160		
22. Tobacco	1633		
23. Cotton lint	1277		
24. Jute and hard fibres	351		

25.	Rubber	564
26.	Beef, veal and buffalo meat	2063
27.	Mutton, lamb and goat meat	2032
28.	Pork	1366
29.	Poultry meat	1300
30.	Milk and dairy products (whole-milk equivalent)	234
31.	Eggs	1120

* Sugar production analysed separately for sugar cane and sugar beet.
† Vegetable oil production analysed separately for soybeans, groundnuts, sesame seed, sunflower seed, all other oils and oilseeds. Demand and trade data and analyses cover oils and oilseeds not produced in the region, e.g. coconuts and palm oil/palm kernel oil.

Appendix 2 Methodology of projections: a summary note

The projections are carried out for each of the commodities and countries analysed individually (see list of commodities and countries covered). The overall quantitative framework for the projections is based on the Supply Utilization Accounts (SUAs). The SUA is an accounting identity showing for any year the sources and uses of agricultural commodities in homogeneous units,[1] as follows:

FOOD + INDUSTRIAL NON-FOOD USES + FEED + SEED + WASTE =
PRODUCTION + (IMPORTS - EXPORTS) + (OPENING STOCKS - CLOSING STOCKS)

There is one such SUA for each of the historical years (generally 1961 to 1986) and the bulk of the projection work is concerned with drawing up SUAs for the year 2000. Different methods are used to project the individual elements of the SUA, as follows.[2]

Food demand per caput was projected using the base year data for this variable (the latest three-year average available at the time the projections were carried out, in this case 1982/4), the FAO food demand model (a set of estimated food demand functions—Engel curves, usually incorporating also trend variables—for up to 52 separate commodities in each country) and assumptions of the growth of per caput incomes (GDP). The results are reviewed and, if considered necessary, modified by commodity and nutrition specialists taking into account any other factors considered relevant and not reflected in the coefficient values for the income elasticities and the time variable. Subsequently, total projected food demand is obtained by simple multiplication of the projected per caput levels with projected population. Prices are not incorporated as an explanatory variable in the food demand function. By implication, price effects are present in the projections only to the extent that they are incorporated into the estimated elasticity and trend coefficients of the periods for which the estimates were

[1] The term Food Balance Sheet (FBS) used in this study is an SUA with the element 'Food' defined at a more detailed level of commodity specification, including the commodities not covered in this study, e.g. animal fats, distilled alcoholic beverages, etc. In the FBS, the element 'Food' is further elaborated to yield levels of food availabilities per caput/year for each commodity. These are subsequently converted into availabilities of calories, proteins and fats per caput/day.
[2] The present text is based in large measure on the methodology description of the FAO study *World Agriculture: Toward 2000* except for the description of the production and trade projections. In that study the projections of production were derived and evaluated by a different and more detailed method than the one described below. For a brief evaluation of the projections methodology see Alexandratos (1988: 277–9).

made. Therefore, from a formal point of view, it is not possible to be more precise as to the extent to which the projections are compatible with any given set of prices, including those of the base year.

The terms 'food demand', 'food consumption' and 'food availability' are used interchangeably to denote the element 'Food' in the SUAs, that is the use of agricultural commodities for human consumption in primary or processed form.

Industrial demand for non-food uses is projected as a function of the GDP growth assumptions and/or the population projections and subsequently adjusted in the process of inspection of the results.

Feed demand for cereals is derived simultaneously with the projections of livestock products by multiplying projected production of each of the livestock products with input/output coefficients for cereals (feeding rates), expressed in terms of metabolizable energy supplied by cereals and brans. These coefficients came originally from the USDA GOL (Grains, Oilseeds, Livestock) Model and have been adjusted over the years as new data and other information became available. The part of total demand met by projected domestic production of brans from the cereals used for food is deducted, and the balance represents feed demand for cereals. Feed use of non-cereal products appearing in the SUAs, e.g. potatoes, is projected by ad hoc methods using historical data mostly as a proportion of total production or total demand. The study did not project systematically feed use of oilmeals except as indicated in Chapter 3.

Seed use is projected as a percentage of production.

Waste is projected as a proportion of total supply (production plus imports).

The study does not project year-2000 *stock changes*. This does not mean that present stocks are assumed to remain constant but rather that, if they are out of balance in the base year, the required changes to adjust them to the levels 'desired' by the market or public authorities are assumed to occur in any year(s) of the projection period, not in the terminal year 2000. The production projected for the year 2000 is, therefore, assumed to be compatible with end-1999 carryover stocks at the 'desired' level.

The *production* projections are, in the first place (i.e. before the adjustments discussed in Ch. 4), extrapolations to year 2000 of the historical trends. These were estimated for each commodity and country or country group for the latest 10– and 15-year periods (in this case 1976-86 and 1971-86) using linear and exponential equations for the variables indicated below, and the best fitting equation was chosen for the extrapolation to year 2000 (subject to the qualifications discussed below).

Linear $Y = a + bt$

Exponential $Y = ae^{gt}$

where Y stands for one of the following variables:

(i) $Y = P_{ij}$ = production of commodity i in country/country group j. This variable was used for the livestock products and for those crops of the study for which separate time series data for area and yield were not available (mostly tree crops).

(ii) $Y = A_j$ = harvested area in country/country group j under all other crops of the study.

(iii) $Y = A_{ij}/A_j$ = share of crop i in total area A.

(iv) $Y = R_{ij}$ = yield per ha of crop i.

The extrapolated production for the year 2000 is then obtained as follows:

(1) If the trends have been negative (b and g negative) the exponential equation is used for the extrapolation while the linear is chosen if trends have been positive. In the former case the choice avoids magnification of the rate of decline over time inherent in linear equations with negative values for b. In the latter case, the linear equation implies an attenuation of past growth trends. When the estimated values of b and g result as statistically non-significant (5 percent level of significance), they are assigned zero values in the extrapolation, i.e. the year 2000 value of the variable extrapolated is equal to that in the base year.

(2) In case (i) above, extrapolated production for year 2000 is

$$P_{ij(2000)} = P_{ij(84/86)} + 15b \text{ (if linear) } OR *e^{15g} \qquad \text{(if exponential)}$$

For the crops with area and yield data,

$$A_{j(2000)} = A_{j(84/86)} + 15b \ OR *e^{15g} \qquad \text{(total area)}$$

$$A_{ij(2000)} = \left[\left(\frac{A_{ij}}{A_j} \right)_{(84/86)} + 15b \ OR *e^{15g} \right] *A_{j(2000)} \qquad \text{(area under crop i)}$$

with $\left(\dfrac{A_{ij}}{A_j} \right)_{(2000)}$ scaled so that $\sum_{i} A_{ij(2000)} = A_{j(2000)}$

$$R_{ij(2000)} = R_{ij(84/86)} + 15b \ OR *e^{15g} \qquad \text{(yield)}$$

and $P_{ii(2000)} = A_{ij(2000)} * R_{ij(2000)}$ \qquad (production)

The extrapolated production is subsequently entered into the SUAs for all commodities in each country together with the projected levels of final demand (food and non-food industrial uses). An internal computing routine fills in the SUA elements of intermediate demand (feed, seed, waste) according to the rules described above for these elements. The implied net trade balances are derived as balancing items and the self-sufficiency ratios computed.

At this stage all countries have complete projected year 2000 SUAs. Those for Eastern Europe and the USSR are adjusted as described in

Chapter 4, i.e. on the basis of *ad hoc* informal judgement. These adjustments lead to levels of production, trade and self-sufficiency which, according to the different specialists who reviewed the projections, represent a feasible combination of policy objectives, technical and economic factors and external conditions.

The resulting net trade balances for Eastern Europe and the USSR together with those of the developing countries (taken from the study *World Agriculture: Toward 2000*) are assumed to represent the external conditions (net trade balances of the rest of the world) facing the Developed Market Economies (DMEs), both those in the study and the four DMEs not covered (i.e. Japan, Australia, New Zealand, South Africa, for which projected SUAs are derived in the same way as for the study countries). The DME group of countries is assumed to act a residual supplier in a global context, given the assumed fixity of the net trade balances of the rest of the world. The resulting confrontation of the net trade balances of the DMEs as a group and the rest of the world provides the basis for the discussion in Chapter 4 of adjustment requirements in the DME country group.

Appendix 3: Statistical tables

Notes to the tables appear in a separate section (pp. 231)

Table A.1 Population and Gross National Product (GNP) per caput

| | Population | | | | GNP per caput | |
| | thousands | | growth rates % p.a | | | dollars | gr. rate 1965–86 |
	1985	2000	1970–80	1980–5	1985–2000	1986	% p.a.
Albania	3050	4102	2.5	2.2	2.0	...	...
Austria	7555	7570	0.1	0.0	0.0	9990	3.3
Belgium	9862	9975	0.2	0.0	0.1	9230	2.7
Bulgaria	8957	9439	0.4	0.2	0.3	...	...
Canada	25380	28874	1.2	1.1	0.9	14120	2.6
Cyprus	669	762	0.2	1.2	0.9	4360	...
Czechoslovakia	15498	16495	0.7	0.3	0.4	...	...
Denmark	5113	5073	0.4	0.0	-0.1	12600	1.9
Finland	4902	5075	0.4	0.5	0.2	12160	3.2
France	55170	57728	0.6	0.5	0.3	10720	2.8
German DR	16650	17041	-0.2	-0.1	0.2	...	...
Germany, FR	61024	59619	0.0	-0.2	-0.2	12080	2.5
Greece	9930	10534	1.0	0.6	0.4	3680	3.3
Hungary	10649	10659	0.4	-0.1	0.0	2020	3.9
Iceland	243	273	1.1	1.3	0.8	13410	3.1
Ireland	3608	4320	1.4	1.2	1.2	5070	1.7
Israel	4233	5348	2.6	1.7	1.6	6210	2.6
Italy	57128	58466	0.5	0.3	0.2	8550	2.6
Luxembourg	363	358	0.7	0.0	-0.1	15770	4.1
Malta	383	418	1.3	0.7	0.6	3450	7.7
Netherlands	14485	15065	0.8	0.5	0.3	10020	1.9
Norway	4150	4223	0.5	0.3	0.1	15400	3.4
Poland	37203	40834	0.9	0.9	0.6	2070	...
Portugal	10212	11211	1.4	0.7	0.6	2250	3.2
Romania	22725	25230	0.9	0.4	0.7	...	...

Spain	38356	42035	1.0	0.6	0.6	4860	2.9
Sweden	8350	8166	0.3	0.0	-0.1	13160	1.6
Switzerland	6520	6486	0.0	0.4	0.0	17680	1.4
Turkey	49289	65351	2.3	2.1	1.9	1110	2.7
United Kingdom	56827	57056	0.1	0.1	0.0	8870	1.7
United States	238816	269163	1.0	1.0	0.8	17480	1.6
USSR	277537	313552	0.9	0.9	0.8	...	...
Yugoslavia	23123	25170	0.9	0.7	0.6	2300	3.9
EC-12	322078	331440	0.5	0.2	0.2	...	...
Other W. Europe	112467	132944	1.4	1.2	1.1	...	...
Total W. Europe	434545	464384	0.7	0.5	0.4	...	...
North America	264196	298037	1.0	1.0	0.8	...	...
W. Europe + N. America	698741	762421	0.8	0.7	0.6	...	...
Eastern Europe	111682	119698	0.6	0.4	0.5	...	...
E. Europe + USSR	389219	433250	0.8	0.8	0.7	...	...
Total Study Countries	1087960	1195671	0.8	0.7	0.6	...	...
Total OECD Countries	807007	878675	0.9	0.7	0.6	...	...
World	4837254	6122101	1.9	1.7	1.6	...	...

Sources: Population 1985 from the UN Population Division supplemented with more recent national estimates (for explanations see FAO, *Production Yearbook 1987*, Rome, 1988); 2000 projections from UN, *World Population Prospects: Estimates and Projections as Assessed in 1984*, Population Studies No. 98, New York, 1986 (medium variant projections). GNP from World Bank, *World Development Report 1988*, Oxford University Press, London, 1988.

Table A.2 Land in agricultural use (1985) and growth rates of agriculture

	Land			Per Caput of		Growth of gross agricultural production (% p.a.)	
	Arable	Permanent crops (thousand ha)	Total	Total Population (ha)	Agricultural labour force (ha)	1961–86	1971–86
Albania	592	121	713	0.23	1.0	3.5	3.4
Austria	1448	77	1525	0.20	6.1	1.2	1.4
Belgium-Luxemb.	790	15	805	0.08	8.3	1.5	0.7
Bulgaria	3810	324	4134	0.46	6.4	2.0	1.4
Canada	45950	80	46030	1.81	86.2	2.3	2.5
Cyprus	103	55	158	0.24	2.2	...	...
Czechoslovakia	5018	135	5153	0.33	5.7	2.1	1.7
Denmark	2617	3	2620	0.51	16.1	1.1	2.2
Finland	2410		2410	0.49	9.8	1.1	1.0
France	17614	1318	18932	0.34	11.3	1.6	1.7
German, DR	4717	256	4973	0.30	5.7	1.9	1.6
Germany, FR	7240	213	7453	0.12	5.6	1.2	1.3
Greece	2900	1040	3940	0.40	3.8	3.1	2.4
Hungary	5036	257	5293	0.50	7.0	2.9	2.5
Iceland	8		8	0.03	0.7	0.9	
Ireland	794	3	797	0.22	3.7	2.5	2.5
Israel	327	91	418	0.10	5.0	3.9	2.2
Italy	9178	3082	12260	0.21	5.8	1.5	1.4
Malta	12	1	13	0.03	2.2	2.2	1.2
Netherlands	863	29	892	0.06	3.4	3.5	3.3
Norway	858		858	0.21	6.3	1.1	1.3
Poland	14511	334	14845	0.40	3.2	1.0	
Portugal	2050	710	2760	0.27	3.0	0.7	0.6
Romania	9985	637	10622	0.47	3.8	3.9	3.8

Spain	15564	4852	20416	0.53	10.9	3.1	2.5
Sweden	2984		2984	0.36	15.1	1.0	1.4
Switzerland	391	21	412	0.06	2.5	1.4	1.5
Turkey	24595	2935	27530	0.56	2.4	3.0	2.8
United Kingdom	7017	60	7077	0.12	11.2	1.7	1.8
United States	187881	2034	189915	0.80	57.0	1.9	1.7
USSR	227800	4387	232187	0.84	10.1	1.9	0.9
Yugoslavia	7046	734	7780	0.34	2.8	2.5	2.2
EC-12	66627	11325	77952	0.24	7.6	1.8	1.8
Other W.Europe	40774	4035	44809	0.40	2.8	2.2	2.2
Total W.Europe	107401	15360	122761	0.28	4.6	1.9	1.9
North America	233831	2114	235945	0.89	61.0	2.0	1.8
W.Eur.+ N.America	341232	17474	358706	0.51	11.8	1.9	1.8
Eastern Europe	43077	1943	45020	0.40	4.2	2.1	1.6
E.Europe + USSR	270877	6330	277207	0.71	8.2	2.0	1.1
Total Study Countries	612109	23804	635913	0.58	9.9	1.9	1.6
Total OECD Countries	386278	17215	403493	0.48	12.5	1.9	1.8
World	1371656	100158	1471814	0.30	1.4	2.5	2.4

Note: *Arable land* refers to land under temporary crops (double-cropped areas are counted only once), temporary meadows for mowing or pasture, land under market and kitchen gardens (including cultivation under glass), and land temporarily fallow or lying idle. *Land under permanent crops* refers to land cultivated with crops that occupy the land for long periods and need not be replanted after each harvest; it includes land under shrubs, fruit trees, nut trees and vines, but excludes land under trees grown for wood or timber.

Table A.3 Per caput food availability for direct human consumption

	Cereals Food	Cereals (feed use)	Potatoes	Sugar (raw equiv.)	Vegetable oils, oilseeds and products (oil equiv.)	Animal fats (product weight)	Meat (carcass weight, excl. offals)	Milk and products (fresh whole milk equiv., excl. butter)	Eggs (fresh equiv.)	Fruit and vegetables	Coffee and cocoa (beans equiv.), Tea	Calories (no.)	Protein (grams)	Fat
					kg. per year							per day		
Albania														
1969/71	230	(32)	37	15	5.5	1.5	25	86	2	100	0.0	2555	76	49
1984/6	242	(76)	34	19	6.0	2.3	26	110	4	88	0.0	2740	83	55
Austria														
1969/71	141	(291)	67	39	10.2	18.5	74	217	14	247	5.0	3233	89	130
1984/6	113	(441)	61	39	13.0	21.8	96	254	14	273	10.6	3416	97	159
Belgium-Lux.														
1969/71	118	(300)	116	41	13.3	21.3	71	182	13	203	8.6	3482	95	164
1984/6	121	(210)	103	40	15.5	31.0	88	203	14	229	9.4	3850	105	195
Bulgaria														
1969/71	255	(372)	28	35	12.8	4.4	40	116	7	286	1.0	3516	97	86
1984/6	220	(527)	31	38	15.1	8.4	67	194	14	258	2.2	3634	106	118
Canada														
1969/71	107	(847)	71	54	11.8	16.9	91	234	15	172	6.4	3250	93	139
1984/6	110	(737)	67	47	20.5	12.7	93	223	12	237	6.6	3425	96	154
Czechoslovakia														
1969/71	181	(397)	104	44	9.2	16.7	73	180	14	134	2.8	3391	98	114
1984/6	172	(482)	80	42	10.1	20.4	90	206	17	164	3.3	3473	103	129
Denmark														
1969/71	110	(1197)	83	52	11.1	31.9	59	250	12	122	14.8	3398	88	162
1984/6	118	(1094)	74	43	12.4	24.7	85	221	15	175	13.5	3512	95	168
Finland														
1969/71	113	(411)	98	50	6.2	21.9	44	322	10	76	12.2	3140	88	128
1984/6	102	(461)	96	38	7.2	18.5	58	324	11	124	13.0	3080	96	130
France														
1969/71	113	(296)	96	39	9.1	14.1	74	234	12	403	6.2	3132	102	109
1984/6	108	(322)	77	39	10.2	20.9	89	266	16	332	7.5	3273	111	136

German DR														
1969/71	152	(355)	151	37	10.8	21.3	73	194	14	131	3.7	3330	94	131
1984/6	162	(611)	147	43	10.3	22.5	109	241	18	181	6.3	3800	113	148
Germany, FR														
1969/71	121	(248)	106	36	12.7	20.5	78	194	16	207	7.7	3210	89	135
1984/6	130	(277)	79	39	14.1	21.5	94	227	16	233	10.3	3475	101	149
Greece														
1969/71	156	(176)	55	22	22.3	1.8	46	156	11	414	2.0	3190	100	116
1984/6	150	(251)	68	34	25.0	2.4	70	220	11	502	3.7	3686	114	151
Hungary														
1969/71	183	(572)	74	37	2.7	26.5	80	147	14	242	3.1	3321	95	111
1984/6	165	(906)	50	38	7.5	31.8	110	162	18	243	6.2	3540	102	142
Iceland														
1969/71	91	(114)	56	54	5.8	12.1	57	375	8	65	12.9	2991	106	112
1984/6	83	(125)	62	52	8.3	13.8	69	325	13	94	14.9	3146	127	135
Ireland														
1969/71	140	(409)	122	55	5.8	17.0	74	269	11	101	5.9	3530	102	139
1984/6	137	(446)	137	46	9.0	20.6	93	167	11	144	5.5	3689	105	149
Israel														
1969/71	146	(245)	36	38	19.4	4.7	49	181	21	388	4.0	3010	97	101
1984/6	135	(261)	34	46	24.2	5.0	52	219	21	368	6.1	3037	99	114
Italy														
1969/71	189	(194)	45	30	19.7	5.1	50	186	11	469	3.7	3399	96	111
1984/6	163	(201)	38	30	20.1	9.4	73	278	11	437	5.6	3494	108	137
Malta														
1969/71	156	(174)	16	38	8.1	6.6	41	156	12	154	5.5	2993	96	97
1984/6	133	(213)	21	40	13.6	4.6	46	184	14	186	3.6	2878	87	110
Netherlands														
1969/71	92	(221)	89	49	20.7	20.2	58	272	13	162	9.3	3203	87	149
1984/6	102	(174)	83	39	24.2	21.8	76	308	14	202	11.3	3258	97	149
Norway														
1969/71	101	(254)	97	44	15.2	18.2	38	258	10	120	11.3	3061	89	135
1984/6	112	(226)	81	41	16.0	16.6	47	298	12	149	13.2	3215	101	139
Poland														
1969/71	200	(358)	145	43	5.4	22.5	53	256	11	137	1.9	3327	100	101
1984/6	187	(440)	109	47	6.4	23.6	62	254	12	159	2.4	3298	102	109
Portugal														
1969/71	140	(97)	113	28	15.6	2.3	32	80	4	337	2.4	2979	84	101
1984/6	162	(182)	94	27	16.9	2.8	43	101	6	285	2.7	3134	91	86

Table A.3 (Cont., page 2) Per caput food availability for direct human consumption

	Cereals Food	Cereals (feed use)	Potatoes	Sugar (raw equiv.)	Vegetable oils, oilseeds and products (oil equiv.)	Animal fats (product weight)	Meat (carcass weight, excl. offals)	Milk and products (fresh whole milk equiv., excl. butter)	Eggs (fresh equiv.)	Fruit and vegetables	Coffee and cocoa (beans equiv.), Tea	Calories (no.)	Protein (grams)	Fat
					kg. per year							per day		
Romania														
1969/71	211	(308)	74	21	8.5	6.9	36	117	7	174	0.6	3055	89	73
1984/6	197	(613)	72	28	9.3	9.4	68	158	15	327	1.0	3358	104	95
Spain														
1969/71	123	(258)	109	29	15.5	2.4	43	119	13	308	2.9	2875	83	98
1984/6	125	(414)	106	29	21.6	2.8	67	162	16	366	4.4	3365	97	142
Sweden														
1969/71	88	(361)	86	45	11.3	18.5	50	287	12	130	14.7	2881	88	117
1984/6	95	(406)	71	46	11.4	20.7	56	370	13	152	13.5	3048	98	126
Switzerland														
1969/71	129	(201)	55	52	14.5	14.3	68	284	11	311	7.5	3494	91	150
1984/6	111	(213)	48	43	14.7	17.4	83	332	12	318	7.2	3431	96	163
Turkey														
1969/71	205	(155)	41	19	9.5	4.1	18	78	2	292	0.8	2851	83	63
1984/6	218	(202)	57	29	15.6	3.0	18	69	4	311	2.8	3148	88	76
United Kingdom														
1969/71	118	(238)	103	52	9.6	18.3	71	227	15	142	7.6	3393	93	148
1984/6	108	(183)	105	40	11.8	14.9	68	225	12	168	7.4	3218	88	143
United States														
1969/71	98	(666)	53	49	17.9	7.8	107	246	18	199	8.6	3392	103	152
1984/6	109	(616)	60	32	23.5	5.1	111	241	15	238	7.2	3643	107	164
USSR														
1969/71	200	(337)	130	42	7.4	10.3	48	192	9	134	1.0	3324	102	85
1984/6	171	(440)	107	47	10.3	15.4	65	171	14	170	1.7	3394	106	102
Yugoslavia														
1969/71	233	(322)	64	30	8.1	10.4	43	119	6	195	2.4	3333	94	84
1984/6	224	(472)	52	35	11.1	13.6	67	168	9	187	1.9	3542	102	110

EC-12														
1969/71	131	(257)	91	38	13.7	13.4	64	197	13	291	6.1	3235	93	125
1984/6	128	(275)	81	36	15.8	15.0	78	232	14	297	7.3	3386	101	143
Other W. Europe														
1969/71	179	(242)	58	31	10.0	10.1	38	155	7	234	4.3	3059	88	91
1984/6	185	(302)	58	34	13.8	9.9	45	167	8	253	5.1	3240	94	103
Total W. Europe														
1969/71	143	(253)	83	36	12.9	12.6	58	187	12	278	5.7	3194	92	117
1984/6	142	(282)	75	36	15.3	13.7	69	216	12	286	6.8	3349	99	133
North America														
1969/71	99	(683)	55	50	17.3	8.6	105	245	18	196	8.4	3378	102	151
1984/6	109	(627)	61	34	23.2	5.8	109	239	15	238	7.1	3622	106	163
W.Eur.+ N.Am.														
1969/71	127	(410)	73	41	14.5	11.2	75	208	14	248	6.6	3261	96	129
1984/6	130	(413)	70	35	18.3	10.7	85	225	13	268	6.9	3452	102	144
Eastern Europe														
1969/71	194	(376)	110	36	7.8	17.3	57	185	11	165	2.1	3298	96	102
1984/6	184	(558)	91	40	8.9	19.7	79	212	15	213	3.2	3460	105	119
E.Europe + USSR														
1969/71	198	(349)	124	40	7.5	12.4	51	190	10	143	1.3	3316	100	90
1984/6	175	(474)	102	45	9.9	16.7	69	183	15	183	2.1	3413	105	107
All Study countries														
All														
1969/71	152	(388)	91	41	12.0	11.6	67	202	12	211	4.7	3281	97	115
1984/6	146	(435)	81	39	15.3	12.9	79	210	14	237	5.2	3438	103	131
OECD countries														
1969/71	126	(364)	68	40	13.6	11.3	68	189	14	238	6.1	3188	94	121
1984/6	128	(368)	64	33	17.7	10.7	78	203	14	254	6.6	3362	100	137
World														
1969/71	147	(120)	32	22	6.8	3.9	26	74	5	116	2.0	2444	65	56
1984/6	166	(129)	28	23	9.2	3.9	30	76	6	128	2.0	2694	70	65

Table A.4 Production: cereals and other major crops (thousand tonnes and growth rates % p.a.)

	Wheat	Rice (paddy)	Maize	Barley	Oats	Rye	Other coarse	Total coarse grains	Total cereals grains	Potatoes	Pulses (dry)	Sugar (raw)	Citrus (fresh)	Tobacco
Albania														
1969/71	242	13	220	8	20	8	27	283	533	109	16	16	4	12
1979/81	492	14	318	24	29	9	30	410	911	112	23	38	10	18
1984/6	562	12	330	35	30	10	35	441	1010	131	24	39	13	20
1987	565	11	320	38	30	10	36	435	1007	135	25	40	15	20
1988	589	11	306	40	30	11	38	424	1020	137	25	45	16	29
1971–86 %	5.6		3.2	12.3	2.6		4.0	3.6	4.6	2.5	4.0	7.7	10.0	3.1
Austria														
1969/71	912		677	954	281	417	99	2429	3341	2787	7	315		1
1979/81	1025		1338	1288	298	327	116	3366	4391	1356	2	450		
1984/6	1493		1670	1444	282	334	119	3848	5341	1054	1	413		1
1987	1451		1685	1179	246	309	98	3517	4968	880	1	390		
1988	1430		1640	1135	250	280	98	3403	4833	901	1	358		
1971–86 %	3.2		6.1	2.5		-1.8	4.0	3.0	3.1	-6.1	-13.6			
Belgium-Lux.														
1969/71	848		11	608	331	82	43	1075	1923	1631	19	711		2
1979/81	949		38	844	171	43	25	1120	2069	1468	7	994		2
1984/6	1290		53	847	130	29	27	1085	2375	1722	7	986		2
1987	1078		40	736	65	20	18	879	1957	1957	19	874		1
1988	1327		40	802	95	14	20	971	2298	2000	16	993		1
1971–86 %	1.9		6.9	1.8	-7.2	-7.5	-4.9				-5.9	2.7		
Bulgaria														
1969/71	2898	64	2436	1109	93	27	22	3686	6627	378	120	218		113
1979/81	3881	71	2627	1439	61	29	22	4178	8107	376	68	165		138
1984/6	4077	59	2398	1074	36	46	19	3573	7689	450	76	90		131
1987	4149	53	1858	1091	41	49	44	3083	7267	316	85	110		133
1988	4713	60	1625	1306	52	58	44	3085	7838	359	93	105		117
1971–86 %	2.1			-2.4	-5.6	6.6				1.8		-6.8		

Canada													
1969/71	13901	2488	10024	5508	474	2053	20546	34447	2313	105	130		105
1979/81	20430	5904	11199	2993	636	1547	22279	42709	2626	186	118		100
1984/6	25606	6553	12411	2854	610	1265	23694	49300	2849	334	96		85
1987	25950	7015	13957	2995	493	1231	25691	51641	3033	892	147		61
1988	15655	5369	10125	2993	257	949	19693	35348	2785	518	109		76
1971–86 %	4.6	7.4		−4.4	4.0	−3.7	1.1	2.7	2.4	7.8	−2.6		
Cyprus													
1979/81	12	74	1				74	87	180	6		109	
1984/6	8	82					83	91	162	5		139	
1987	14	112	1				113	126	150	5		153	
1988	13	130	1				131	144	195	4		135	
Czechoslovakia													
1969/71	3436	511	2543	882	586	58	4581	8017	4864	96	731		6
1979/81	4482	800	3524	418	534	4	5279	9762	3388	137	808		5
1984/6	5833	1015	3582	457	625	9	5688	11521	3647	221	857		6
1987	6154	1160	3551	406	496	9	5622	11777	3072	224	818		6
1988	6547	996	3411	366	534	8	5314	11861	3659	253	614		5
1971–86 %	2.1	3.9	1.7	−3.8		−9.2	1.1	1.6	−2.4	9.1	1.2		
Denmark													
1969/71	509	5175	699		136	158	6169	6678	845	72	306		
1979/81	692	6250	166		221	16	6653	7345	913	14	493		
1984/6	2198	5486	146		573	15	6219	8417	1117	460	572		
1987	2285	4292	95		513	14	4913	7198	957	524	422		
1988	2080	5419	202		366	25	6012	8092	942	579	550		
1971–86 %	10.6		−11.2		9.0	−15.2	1.1	1.3	3.0	21.6	3.3		
Finland													
1969/71	445	943	1297		130	62	2431	2875	906	3	65		
1979/81	267	1421	1183		88	34	2726	2993	629	13	104		
1984/6	493	1761	1238		78	33	3110	3603	742	10	121		
1987	281	1306	813		74	15	2208	2489	491	4	64		
1988	285	1612	857		49	23	2541	2826	855	4	147		
1971–86 %		3.8			−4.3	−4.5	1.5			5.9	3.3		

Table A.4 (cont., page 2) Production: cereals and other major crops (thousand tonnes and growth rates % p.a.)

	Wheat	Rice (paddy)	Maize	Barley	Oats	Rye	Other coarse grains	Total coarse grains	Total cereals	Potatoes	Pulses (dry)	Sugar (raw)	Citrus (fresh)	Tobacco
France														
1969/71	14112	88	7394	8865	2317	297	776	19650	33820	8569	135	2873	12	45
1979/81	22362	25	9641	10997	1850	368	848	23704	46082	6735	340	4720	34	47
1984/6	29412	52	11514	11005	1570	292	1060	25441	54888	7006	1064	4121	33	36
1987	27415	54	12470	10489	1122	299	1116	25496	52946	6720	1956	3973	37	38
1988	29677	65	13996	10086	1074	276	1004	26435	56156	6344	2623	4424	30	33
1971–86 %	4.5		2.3		-3.2				2.6		20.0	2.8	5.9	-2.4
German DR														
1969/71	2203		9	2093	735	1594	406	4836	7039	10432	87	487		5
1979/81	3052		4	3592	571	1848	48	6064	9115	10612	82	675		4
1984/6	4011		1	4266	704	2474	95	7540	11551	11418	100	789		6
1987	4040		1	4198	637	2283	60	7179	11219	12228	90	760		6
1988	3697		1	3798	508	1783	30	6119	9816	11473	91	445		6
1971–86 %	2.6		-19.5	3.3		2.5	-8.9	2.0	2.2		1.6	2.3		
Germany, FR														
1969/71	6268		500	5219	2832	2862	1376	12790	19058	15804	95	2158		9
1979/81	8177		748	8566	2777	1980	681	14752	22928	8853	35	3261		8
1984/6	10165		1177	9784	2530	1840	502	15833	25998	8294	148	3362		8
1987	9932		1217	8571	2008	1599	437	13832	23764	7354	382	2963		6
1988	12044		1435	9609	2036	1558	449	15087	27131	7353	403	3130		8
1971–86 %	2.8		6.3	3.7	-4.3	-3.6	-7.5	0.9	1.6	-4.9		2.9		-2.5
Greece														
1969/71	1867	84	498	655	108	9	10	1280	3203	700	123	163	585	88
1979/81	2770	84	1165	838	80	7	6	2097	4923	1036	94	286	802	125
1984/6	2166	107	2022	708	67	23	2	2821	5063	1017	63	306	1030	152
1987	2213	137	2156	573	70	32	2	2833	5138	948	54	197	776	155
1988	2550	116	2116	695	74	31	2	2918	5545	850	54	254	1013	142
1971–86 %	1.6		12.0	-1.8	-4.3	8.0	-12.5	5.2	3.4	2.9	-4.5	4.9	3.5	3.7

	1	2	3	4	5	6	7	8	9	10	11	12	13	14
Hungary														
1969/71	3410	54	4542	749	79	193	30	5593	9039	1874	138	321		20
1979/81	4800	35	7022	848	125	117	55	8166	12989	1504	127	511		21
1984/6	6588	39	6921	1041	138	177	42	8320	14934	1398	230	475		21
1987	5748	40	7234	794	99	186	69	8381	14155	1077	274	493		20
1988	6962	52	6027	1161	134	245	54	7621	14618	1128	270	521		23
1971–86 %	3.2	–3.4	2.0	1.9	5.0			1.9	2.4	–1.3	4.6	3.8		1.6
Iceland														
1969/71										7				
1979/81										11				
1984/6										12				
1987										11				
1988										11				
Ireland														
1969/71	375			854	222	1		1077	1452	1450	3	164		
1979/81	268			1603	95	1		1699	1967	902	179			
1984/6	507			1564	116	1		1681	2188	725	1	210		
1987	402			1599	106	1		1706	2108	697	1	242		
1988	418			1538	117	1		1656	2074	680	1	212		
1971–86 %	4.7			4.3	–4.3	–1.5		3.4	3.6	–4.7		1.4		
Israel														
1969/71	160		5	17	1		16	39	199	131	8	28	1318	2
1979/81	201		13	18			6	38	239	200	8	8	1518	1
1984/6	142		22	9			6	37	179	203	8		1435	1
1987	298		25	21			1	46	344	218	11		1502	1
1988	211		25	20			1	46	257	207	11		1282	1
1971–86 %	–3.2		6.2				–11.3	3.4	–3.1	2.5				–2.9
Italy														
1969/71	9756	858	4601	326	488	65	21	5501	15829	3633	613	1271	2502	79
1979/81	8989	989	6590	914	433	35	75	8047	17695	2941	321	1956	2798	131
1984/6	9207	1090	6497	1597	396	23	95	8608	18542	2466	272	1535	3235	158
1987	9381	1064	5764	1710	361	20	99	7955	18045	2454	233	1867	2314	162
1988	7945	1094	6318	1561	383	18	105	8384	17059	2330	213	1607	3187	161
1971–86 %	–3.2	1.5	2.7	10.4	–1.5	–4.2	19.6	3.4	1.3	–1.5	–4.6	1.9	1.6	4.8

Table A.4 (cont., page 3) Production: cereals and other major crops (thousand tonnes and growth rates % p.a.)

	Wheat	Rice (paddy)	Maize	Barley	Oats	Rye	Other coarse grains	Total coarse grains	Total cereals	Potatoes	Pulses (dry)	Sugar (raw)	Citrus (fresh)	Tobacco
Malta														
1969/71	2			2				2	4	20	1		1	
1979/81	4			3				3	8	21	2		1	
1984/6	5			5				5	10	13	2		1	
1987	5			5				5	10	13	2		1	
1988	5			5				5	10	13	2		1	
1971–86 %	8.4			8.9				8.7	8.7	-3.3			6.3	
Netherlands														
1969/71	675		5	366	243	196	12	822	1497	5367	46	769		
1979/81	867		3	265	106	39	1	413	1280	6329	24	1000		
1984/6	974		4	217	52	21	1	295	1269	6893	109	1105		
1987	769		5	262	47	25		339	1107	7478	173	1064		
1988	816		5	311	63	27		406	1222	6742	106	1075		
1971–86 %	3.2		-11.0	-4.0	-7.8	-13.8	-47.8	-6.2		1.4	9.0	2.4		
Norway														
1969/71	11			545	215	5	1	766	777	807				
1979/81	63			636	424	4	2	1066	1129	524				
1984/6	166			609	492	4		1106	1272	442				
1987	249			567	466	4		1037	1286	371				
1988	200			600	480	3	2	1085	1285	414				
1971–86 %	16.6			2.0	4.0	-4.5	3.4	2.1	2.9	-3.1				
Poland														
1969/71	4924		12	2182	3156	7143	819	13311	18236	45013	350	1582		82
1979/81	4189		102	3563	2387	6166	2059	14277	18466	39508	216	1530		75
1984/6	6658		80	4018	2591	8072	2973	17732	24390	37674	447	1860		112
1987	7942		146	4335	2429	6817	4393	18118	26060	36252	512	1823		114
1988	7582		204	3804	2222	5501	5191	16922	24504	34707	565	1824		91
1971–86 %			11.4	2.0	-1.8		7.5	1.1	0.9	-2.2				2.6

Portugal														
1969/71	614	177	599	65	92	164	9	929	1661	1222	113	12	138	10
1979/81	335	137	486	45	79	128	10	748	1174	1103	77	6	141	7
1984/6	456	143	566	82	141	100	10	899	1451	1147	80	6	140	5
1987	534	144	655	80	155	108	10	1008	1638	1178	80	2	148	4
1988	401	151	663	48	76	73	10	870	1371	795	78	2	147	4
1971–86 %	−3.4	6.0	5.0	9.4		−3.4	−4.0	5.5	4.1	6.0	−2.4	−6.5		−4.7
Romania														
1969/71	4433	67	7354	615	138	52	3	8162	12639	2671	214	434		26
1979/81	5471	49	11823	2360	57	38	28	14307	19811	4381	116	595		35
1984/6	6855	139	16224	2265	115	53	15	18671	25619	7597	316	585		33
1987	9672	154	18378	3231	100	55	15	21779	31553	7572	432	450		27
1988	9000	150	19500	2200	160	60	20	21940	31040	8000	337	430		32
1971–86 %														−1.9
Spain														
1969/71	4734	384	1804	3922	504	292	181	6702	11693	4985	594	888	2334	24
1979/81	4510	435	2227	6571	527	239	189	9754	14554	5615	365	934	2853	38
1984/6	5258	466	3122	9640	634	269	109	13773	19342	5678	329	1084	3305	41
1987	5791	483	3557	9836	502	318	88	14301	20414	5552	335	1108	4528	36
1988	6514	499	3577	12070	537	357	106	16647	23494	4578	316	1088	4047	40
1971–86 %		6.0	3.1	4.2			−4.0	3.5	2.5	6.0	−4.7	2.3	1.7	4.4
Sweden														
1969/71	958			1836	1561	239	196	3831	4789	1221	14	233		
1979/81	1088			2323	1635	197	163	4319	5407	1191	32	350		
1984/6	1615			2456	1686	186	169	4498	6112	1323	125	375		
1987	1558			1907	1440	137	128	3613	5171	1068	125	275		
1988	1357			1942	1402	140	111	3595	4952	1241	125	396		
1971–86 %				1.8		−5.6		3.5	0.9	0.9	16.9	2.3		
Switzerland														
1969/71	368		61	147	36	48	12	304	672	1016		65		2
1979/81	409		121	220	53	35	5	434	843	924	1	119		2
1984/6	546		152	271	45	22	2	492	1038	828	1	133		2
1987	462		144	241	38	18	2	444	905	658	1	123		1
1988	553		237	299	48	19	3	606	1159	748	1	150		2
1971–86 %	2.5		3.3	3.5		−5.9	−13.7	2.1	2.3	2.3		5.6		

Table A.4 (cont., page 4) Production: cereals and other major crops (thousand tonnes and growth rates % p.a.)

	Wheat	Rice (paddy)	Maize	Barley	Oats	Rye	Other coarse grains	Total coarse grains	Total cereals	Potatoes	Pulses (dry)	Sugar (raw)	Citrus (fresh)	Tobacco
Turkey														
1969/71	11423	257	1058	3720	446	781	346	6351	17945	1984	591	700	648	157
1979/81	17058	314	1263	5480	350	558	208	7860	25127	2957	818	1178	1158	204
1984/6	17766	275	1900	6667	310	357	114	9347	27297	3767	1572	1489	1238	169
1987	18932	275	2400	6900	325	380	70	10075	29191	4300	2134	1784	1343	185
1988	20500	263	2100	7500	276	293	53	10222	30897	4350	2298	1595	1474	212
1971–86 %	3.0		3.6	4.8	-2.2	-5.8	-7.9	3.0	3.0	3.9	7.3	5.2	4.2	
United Kingdom														
1969/71	4140		2	8257	1300	14	228	9801	13941	7359	252	1033		
1979/81	8116		1	10058	587	25	53	10724	18840	6601	240	1215		
1984/6	13642			10273	545	32	32	10881	24523	6912	398	1393		
1987	11941			9226	451	32	28	9737	21678	6760	559	1335		
1988	11605		1	8765	557	33	22	9378	20983	6812	588	1413		
1971–86 %	8.5		-68.4	1.4	-6.2	4.3	-13.6	3.0	3.9	1.1	3.0	3.7		
United States														
1969/71	40034	3953	122824	9476	13350	984	19330	165964	208635	14480	1038	5222	10431	819
1979/81	66229	6968	192366	8838	7234	474	19183	228095	298971	14876	1457	5345	13583	813
1984/6	64514	6155	210337	13071	6681	615	24792	255495	324115	17104	1212	5637	9789	666
1987	57357	5879	179988	11529	5429	503	18807	216257	277535	17484	1463	6651	10857	540
1988	49295	7237	125353	6325	3175	382	14700	149935	204057	15875	1140	6260	11547	604
1971–86 %	3.1	3.1	2.7	3.0	-3.5			2.2	2.4	1.1	2.1			-1.8
USSR														
1969/71	92804	1272	9993	35132	13974	12235	3911	75244	168896	93739	7468	8722	72	256
1979/81	89859	2558	9076	42502	14372	9309	2780	78039	169604	76706	5055	7017	261	285
1984/6	79672	2640	13486	47426	20533	14997	3768	100209	181642	81903	8767	8548	274	378
1987	83312	2683	14808	58409	18495	18082	5700	115494	200595	75908	9898	9565	184	303
1988	84500	2900	16000	47000	16500	16000	4160	99660	186094	62700	8735	9240	200	340
1971–86 %	3.1	3.9	2.3	3.0	2.0			2.2	2.4	1.1	2.1	1.7	13.2	1.7

Yugoslavia														
1969/71	4760	32	7399	442	310	132	42	8324	13105	3020	285	454	1	46
1979/81	4625	40	9736	726	296	78	21	10857	15508	2646	204	798	5	65
1984/6	5077	40	11246	719	256	78	24	12322	17426	2513	197	937	6	85
1987	5345	49	8863	504	232	69	14	9682	15060	2210	190	872	6	76
1988	6303	36	7697	616	253	76	15	8657	14983	1935	163	636	9	52
1971–86 %	3.6	2.1	2.7	1.7	-1.5	-3.6	-3.7	2.5	1.5		-1.9	5.9	9.9	2.7
EC-12														
1969/71	43897	1590	15415	34312	9137	4117	2816	65797	110753	51565	2064	10348	5572	256
1979/81	58034	1670	20900	46951	6871	3086	1904	79710	138858	42496	1516	15044	6628	358
1984/6	75280	1858	24956	51201	6327	3203	1852	87536	164055	42976	2933	14680	7742	402
1987	71740	1883	25864	47374	4982	2967	1813	82999	155994	42055	4317	14048	7802	402
1988	75377	1924	28151	50904	5213	2754	1743	88764	165424	39426	4978	14748	8423	389
1971–86 %	3.1	1.1	3.1	2.0	-3.0	-2.2	-3.7	1.4	2.3	-1.2	3.1	2.7	1.8	3.0
Other W. Europe														
1969/71	19354	302	9421	8703	4168	1759	800	24851	44406	12191	937	1875	2177	220
1979/81	25242	368	12790	12213	4269	1297	586	31155	56642	10752	1107	3045	2801	291
1984/6	27873	327	15320	14057	4339	1070	502	35288	63379	11190	1946	3507	2832	277
1987	29159	335	13437	12780	3591	1002	365	31174	60556	10504	2497	3548	3020	283
1988	31446	309	12006	13899	3597	870	344	30715	62367	11006	2635	3326	2916	295
1971–86 %	2.1	1.3	3.2	3.4		-4.4	-3.6	2.3	2.2		5.6	4.3	0.9	
Total W.Europe														
1969/71	63251	1892	24836	43015	13305	5876	3616	90647	155158	63756	3002	12223	7749	476
1979/81	83276	2038	33690	59164	11140	4383	2489	110864	195499	53247	2623	18089	9428	649
1984/6	103152	2185	40276	65258	10666	4272	2354	122824	227433	54166	4878	18187	10574	679
1987	100899	2217	39300	60154	8573	3968	2178	114172	216549	52559	6814	17596	10822	685
1988	106822	2233	40157	64803	8810	3624	2086	119479	227790	50432	7612	18074	11339	684
1971–86 %	3.1	1.2	3.1	2.3	-1.9	-2.8	-3.7	1.7	2.3	-1.0	4.1	3.0	1.6	1.9
North America														
1969/71	53935	3953	125311	19500	18858	1458	21383	186510	243082	16793	1143	5352	10431	924
1979/81	86659	6968	198270	20037	10227	1111	20730	250374	341681	17502	1643	5463	13583	913
1984/6	90120	6155	216890	25483	9535	1225	26057	279189	373415	19953	1545	5734	9789	750
1987	83307	5879	187003	25486	8424	996	20038	241947	329175	20517	2355	6798	10857	601
1988	64950	7237	130722	16450	6168	639	15649	169628	239405	18659	1658	6369	11547	680
1971–86 %	3.5	3.1	2.8	2.0	-3.7			2.1	2.5	1.3		3.0		-1.7

Table A.4 (cont., page 5) Production: cereals and other major crops (thousand tonnes and growth rates % p.a.)

	Wheat	Rice (paddy)	Maize	Barley	Oats	Rye	Other coarse grains	Total coarse grains	Total cereals	Potatoes	Pulses (dry)	Sugar (raw)	Citrus (fresh)	Tobacco
W.Eur.+ N.Am.														
1969/71	117185	5846	150147	62515	32163	7335	24999	277156	398240	80548	4145	17575	18180	1400
1979/81	169934	9006	231960	79201	21366	5494	23219	361238	537180	70749	4265	23552	23011	1561
1984/6	193272	8340	257165	90740	20201	5497	28410	402013	600847	74119	6423	23920	20363	1429
1987	184206	8096	226303	85639	16997	4964	22216	356118	545724	73076	9169	24394	21679	1287
1988	171772	9470	170879	81253	14979	4263	17735	289107	467195	69091	9270	24444	22885	1364
1971–86 %	3.3	2.6	2.9	2.2	-2.8	-2.1	5.3	2.0	2.4		3.9	2.2		
Eastern Europe														
1969/71	21305	185	14863	9292	5083	9594	1338	40169	61598	65232	1004	3772		252
1979/81	25875	155	22378	15327	3618	8732	2217	52271	78249	59769	746	4284		279
1984/6	34021	238	26638	16245	4041	11447	3154	61525	95704	62183	1389	4655		308
1987	37704	247	28775	17200	3711	9885	4589	64161	102029	60517	1617	4454		305
1988	38501	262	28353	15680	3441	8180	5347	61001	99677	59325	1608	3939		274
1971–86 %	1.8		3.5	2.6	-1.9			2.2	2.1	-1.2	3.4	0.9		
E. Europe+USSR														
1969/71	114109	1456	24856	44423	19057	21828	5248	115413	230493	158971	8472	12495	72	509
1979/81	115735	2714	31453	57829	17991	18041	4996	130309	247853	136475	5801	11300	261	564
1984/6	113693	2878	40124	63671	24574	26443	6922	161733	277346	144086	10156	13204	274	685
1987	121015	2930	43583	75609	22206	27967	10289	179655	302624	136424	11515	14019	184	608
1988	123001	3162	44353	62680	19941	24180	9507	160660	285770	122025	10343	13179	200	614
1971–86 %		3.7	3.0		1.1	1.9	1.9	1.5		-1.1			13.2	1.0
All Study countries														
1969/71	231294	7302	175003	106938	51220	29163	30247	392569	628733	239519	12616	30069	18252	1908
1979/81	285669	11720	263413	137029	39357	23534	28215	491547	785033	207224	10067	34853	23273	2125
1984/6	306966	11218	297289	154411	44774	31940	35332	563746	878193	218205	16580	37124	20636	2114
1987	305221	11026	269886	161248	39203	32931	32505	535773	848348	209501	20684	38412	21863	1895
1988	294773	12632	215231	143933	34920	28444	27241	449767	752966	191116	19613	37623	23085	1978
1971–86 %	1.7	2.8	2.9	1.6	-0.9	1.9	1.9	1.9	1.8	-0.9	1.9	1.5		

All OECD countries														
1969/71	121876	22348	142809	65180	33328	7214	25707	274236	411018	81641	4126	20216	19781	1519
1979/81	179949	22951	222223	82280	22492	5419	24339	356752	552008	71938	4334	26741	25649	1639
1984/6	205627	23688	246010	95510	21410	5441	30185	398555	619982	76206	7115	27320	22327	1460
1987	191467	22014	217478	89130	18517	4906	23867	353898	560047	75601	10659	27874	23851	1307
1988	180009	22583	163249	84516	16593	4206	19623	288187	483258	71954	10787	28277	25240	1400
1971–86 %	3.6	3.0	2.9	2.3	−2.5	−2.0		2.0	2.5		4.5	2.1		2.7
World														
1969/71	327890	310065	283487	125190	54494	30781	100616	594563	1129229	285764	43116	70193	38123	4612
1979/81	443606	396226	422075	156741	42582	24980	103889	750262	1458108	269696	40799	88466	56501	5560
1984/6	519953	471742	475420	177151	48233	33284	111670	845755	1680308	288612	50795	99680	59873	6511
1987	517131	464427	458291	181689	43223	34145	103992	821336	1648200	283767	54612	101693	63613	6200
1988	509932	483385	405728	168413	38848	29617	106853	749456	1581774	269619	54645	102687	67096	6528
1971–86 %	3.0	3.0	3.1	1.6	−0.8	0.6	0.8	2.0	2.5	1.2	2.5	2.7		2.2

Table A.5 Production: major oilseeds, cotton, livestock products (thousand tonnes and growth rates % p.a.)

	Soybeans	Sunflower seed	Rapeseed	Olive oil	Cotton seed	Cotton lint	Beef	Mutton	Pigmeat	Poultry-meat	Total Meat	Milk	Eggs
							carcass weight, excl. offals						
Albania													
1969/71		18		4	10	5	18	24	8	5	54	226	4
1979/81		33		5	14	8	25	26	9	10	70	395	10
1984/6		38		4	9	5	26	26	9	13	75	420	13
1987		30		4	9	4	27	26	9	14	76	423	13
1988		27		4	9	4	27	27	9	15	78	425	14
1971–86 %		5.6		-3.3			2.7	1.1	1.2	7.3	2.5	4.0	9.4
Austria													
1969/71			8				191	2	331	47	572	3347	86
1979/81			7				234	4	417	72	727	3449	97
1984/6			19				257	4	453	81	794	3790	100
1987			66				254	4	446	84	788	3734	101
1988			81				244	4	451	87	786	3639	103
1971–86 %			9.3				1.7	6.1	2.6	3.4	2.4	1.1	1.1
Belgium-Luxemb.													
1969/71			1				252	3	530	116	901	3971	233
1979/81			1				305	4	673	133	1115	4042	201
1984/6			9				340	4	683	158	1185	4144	170
1987			18				349	5	730	172	1256	4056	161
1988			14				350	5	735	179	1269	3900	160
1971–86 %			15.6				1.9	2.2	0.6	2.1	1.1	0.6	-2.8
Bulgaria													
1969/71	8	471			26	13	90	86	160	94	430	1628	92
1979/81	123	421			9	5	122	90	318	152	682	2240	134
1984/6	54	438			10	5	134	106	347	164	750	2576	156
1987	33	410			13	7	132	98	372	169	771	2590	161
1988	18	367			9	5	126	98	410	178	812	2586	162
1971–86 %	8.2				-6.0	-6.1	3.1	1.6	4.9	2.9	3.6	3.5	3.9

	1	2	3	4	5	6	7	8	9	10
Canada										
1969/71	257	39	1517	888	7	604	426	1925	8247	328
1979/81	651	183	2581	1020	5	822	532	2379	7830	331
1984/6	963	61	3565	1088	8	954	582	2633	7971	325
1987	1270	52	3847	1028	8	964	653	2653	7986	324
1988	1153	60	4243	1031	8	998	670	2707	8150	323
1971–86 %	9.3	20.7	7.0	1.0		3.6	2.4	2.1		0.3
Cyprus										
1979/81			2	2	5	16	10	34	72	5
1984/6			2	3	7	23	13	47	107	6
1987			1	4	9	25	14	52	118	7
1988			2	4	9	26	13	52	123	7
Czechoslovakia										
1969/71		4	70	325	8	557	110	1000	5009	216
1979/81		28	165	383	7	828	192	1409	5887	261
1984/6		47	297	423	12	819	197	1451	6947	294
1987		62	337	430	12	835	220	1497	6982	295
1988		63	380	443	13	851	242	1548	7024	298
1971–86 %			8.4	1.4	3.1	2.1	3.4	2.1	2.0	1.7
Denmark										
1969/71			30	225	2	743	76	1046	4586	84
1979/81			204	248	1	964	100	1312	5126	77
1984/6			545	242	1	1090	113	1447	5148	80
1987			562	235	1	1150	112	1498	4860	77
1988			528	216		1119	114	1451	4728	75
1971–86 %			18.1	1.1		3.0	2.2	2.6	0.9	0.9
Finland										
1969/71			9	109	1	109	8	227	3401	64
1979/81			68	115	1	171	15	303	3236	78
1984/6			100	125	1	172	21	320	3126	87
1987			90	124	1	176	27	328	2938	81
1988			121	112	1	169	28	310	2753	77
1971–86 %			19.6	1.2		2.9	6.9	2.4	–0.2	0.9

Table A.5 (cont., page 2) Production: major oilseeds, cotton, livestock products (thousand tonnes and growth rates % p.a.)

	Soybeans	Sunflower seed	Rapeseed	Olive oil	Cotton seed	Cotton lint	Beef	Mutton	Pigmeat	Poultry-meat	Total Meat	Milk	Eggs
							carcass weight, excl. offals						
France													
1969/71	19	55	603	3			1640	120	1312	630	3702	28514	645
1979/81		280	871	2			1969	168	1660	1131	4927	33592	851
1984/6	66	1454	1275	1			2077	170	1579	1281	5107	33202	920
1987	210	2659	2655	2			2100	163	1667	1384	5314	30107	873
1988	255	2457	2470	2			2153	164	1653	1429	5399	29012	913
1971–86 %		26.8	5.7	2.5			1.6	2.6	0.7	5.0	2.0	1.0	2.5
German DR													
1969/71			181				350	12	843	89	1293	7243	258
1979/81			264				399	24	1275	145	1843	8260	323
1984/6			377				432	22	1371	158	1984	9060	335
1987			366				472	26	1440	160	2098	9252	335
1988			424				457	30	1467	161	2115	9219	335
1971–86 %			2.6				0.9	4.2	2.7	3.0	2.3	1.1	1.8
Germany, FR													
1969/71			191				1280	13	2614	251	4159	21783	884
1979/81			354				1508	21	3103	386	5018	24537	801
1984/6			811				1657	24	3200	373	5254	26081	780
1987			1265				1703	27	3286	399	5415	24458	739
1988			1181				1600	33	3242	407	5283	24000	726
1971–86 %			10.1				2.0	2.8	1.6	2.8	1.8	1.8	−1.4
Greece													
1969/71		2		179	224	111	83	86	54	69	291	1353	100
1979/81		4		275	232	115	101	123	144	138	505	1676	123
1984/6		109		294	337	170	76	121	146	155	497	1680	122
1987		177		248	345	176	70	124	164	150	507	1697	117
1988		60		260	452	222	69	124	160	149	502	1704	124
1971–86 %		29.6		2.5	1.9	2.1	1.7	1.7	5.2	4.0	2.6	0.9	1.3

Country / Year	(1)	(2)	(3)	(4)	(5)	(6)	(7)	(8)	(9)	(10)	(11)	(12)	(13)
Hungary													
1969/71	178	1703	1010	215	597	19	180			46	122	38	
1979/81	253	2567	1490	346	939	21	184			71	500	49	
1984/6	241	2768	1691	415	1077	24	175			100	713	69	
1987	239	2827	1718	463	1072	20	163			108	803	100	
1988	204	2837	1681	482	1027	20	152			80	640		
1971–86 %	1.9	4.1	3.0	4.6	3.0	3.0	2.7			3.9	14.9		
Iceland													
1969/71	2	117	15	1	1	12	2						
1979/81	3	121	20	1	2	15	3						
1984/6	3	123	19	2	2	13	2						
1987	4	119	20	2	2	14	3						
1988	4	114	24	2	2	18	2						
1971–86 %	3.0	−0.4	1.1	10.2	10.3	3.0							
Ireland													
1969/71	40	3685	549	36	146	42	325		1				
1979/81	35	5392	687	49	135	40	463			1			
1984/6	37	6196	774	55	133	46	540			12			
1987	35	5751	789	62	153	52	523			16			
1988	34	5463	761	61	145	55	501			10			
1971–86 %	−0.7	3.7	1.8	2.5	4.2	3.4	2.7						
Israel													
1969/71	75	495	107	80	5	4	19	37	60		4		
1979/81	92	747	187	153	9	4	20	83	137		8		
1984/6	111	871	199	164	11	5	19	85	139		8		
1987	99	952	195	157	12	5	22	59	98		7		
1988	101	935	195	157	12	5	22	74	122		6		
1971–86 %	2.2	3.5	3.4	3.8	4.2	3.4	2.1	6.6	6.7		5.2		
Italy													
1969/71	580	10081	2032	619	568	36	808	1	2	5	9	401	504
1979/81	659	11336	2954	1007	1005	54	889		1	1	66	1589	591
1984/6	635	11710	3135	1014	1113	52	957			21	188	1393	447
1987	644	11696	3122	1023	1116	50	934			68	235		658
1988	706	11692	3123	1047	1135	47	894			56	257		430
1971–86 %	1.2	1.2	3.2	2.8	4.7	4.6	2.1				17.9		

Table A.5 (cont., page 3) Production: major oilseeds, cotton, livestock products (thousand tonnes and growth rates % p.a.)

	Soybeans	Sunflower seed	Rapeseed	Olive oil	Cotton seed	Cotton lint	Beef	Mutton	Pigmeat	Poultry-meat	Total Meat	Milk	Eggs
							carcass weight, excl. offals						
Malta													
1969/71							1		4	3	7	28	5
1979/81							2			5	7	31	6
1984/6							1		6	4	12	30	7
1987							1		7	4	12	30	7
1988							1		7	4	13	31	7
1971–86 %							2.4			2.7			1.1
Netherlands													
1969/71			22				336	10	743	285	1374	8182	268
1979/81			28				431	23	1350	343	2146	11832	540
1984/6			30				519	18	1652	390	2579	12676	662
1987			31				519	17	1828	416	2781	11667	597
1988			24				525	19	1877	442	2862	11315	600
1971–86 %							3.1	3.6	5.2	1.6	4.1	2.9	8.1
Norway													
1969/71			6				58	17	66	6	147	1752	38
1979/81			10				73	20	83	12	188	1952	45
1984/6			14				73	25	84	12	195	1994	52
1987			11				79	26	92	15	212	2016	52
1988			12				79	25	98	12	214	1982	57
1971–86 %			10.4				1.9	3.8	1.0	3.6	1.8	0.9	2.6
Poland													
1969/71			455				552	26	1300	130	2007	15022	387
1979/81			434				688	26	1615	417	2745	16329	489
1984/6			1094				756	34	1497	282	2569	16400	466
1987			1186				786	42	1714	325	2867	15615	443
1988			1199				776	42	1567	380	2765	15490	445
1971–86 %			3.2				1.8	1.7	1.0	3.6	3.6		0.7

Portugal													
1969/71	102	1		66			87	22	98	54	260	721	37
1979/81	366	14		41			106	23	164	131	425	870	62
1984/6	399	30		45			99	25	179	112	414	968	66
1987	633	29		39			102	26	208	134	470	1135	73
1988	400	51		20			92	25	190	135	442	1033	74
1971–86 %	5.1	22.3					2.0	0.8	4.4	3.9	3.3	2.2	4.6
Romania													
1969/71	2	769	3				240	81	457	127	905	3141	184
1979/81	12	838	15				315	91	971	413	1790	4385	350
1984/6	4	855	44		1		293	90	985	476	1843	4622	422
1987	4	1102	38		1		267	82	892	425	1666	4730	506
1988	4	1190	40		1		257	86	842	390	1576	4760	425
1971–86 %	5.1	10.1	14.5				2.0		4.0	7.7	4.0	1.7	4.9
Spain													
1969/71		146	15	379	94	52	294	136	468	454	1351	5000	465
1979/81		515	9	393	95	58	405	165	1115	798	2483	6497	668
1984/6		962	10	531	110	70	401	208	1377	789	2776	6913	661
1987		1006	13	734	137	79	439	224	1414	797	2874	6655	726
1988		1123		357	198	118	437	234	1590	826	3087	7230	759
1971–86 %		10.1				2.8	1.2	3.2	8.5	3.2	4.9	2.2	2.5
Sweden													
1969/71			218				158	3	244	27	432	2996	102
1979/81			313				156	5	317	44	522	3452	114
1984/6			371				154	5	322	47	527	3662	120
1987			284				140	5	291	44	480	3464	119
1988			305				135	5	295	42	477	3429	117
1971–86 %			1.6				1.2	3.4	1.5	3.0	1.6	1.7	1.0
Switzerland													
1969/71			19				134	4	201	16	354	3193	39
1979/81			33				163	4	272	24	463	3677	43
1984/6			41				169	4	282	27	482	3869	43
1987			49				178	5	281	29	492	3783	43
1988			50				175	5	285	29	494	3813	44
1971–86 %			4.8				1.8	2.0	2.0	2.6	2.0	1.4	0.5

Table A.5 (cont., page 4) Production: major oilseeds, cotton, livestock products (thousand tonnes and growth rates % p.a.)

	Soybeans	Sunflower seed	Rapeseed	Olive oil	Cotton seed	Cotton lint	Beef	Mutton	Pigmeat	Poultry-meat	Total Meat	Milk	Eggs
							carcass weight, excl. offals			excl. offals			
Turkey													
1969/71	11	383	5	75	705	441	202	334	1	104	640	4338	98
1979/81	7	638	20	113	781	488	234	323	1	245	802	5497	217
1984/6	128	817		121	862	539	253	414	1	271	938	4952	282
1987	250	1100		90	859	537	254	416		284	954	5024	306
1988	150	1150		150	923	577	258	413		290	962	4924	306
1971–86 %	19.9	3.9					2.4	1.8		6.0	2.6	1.0	7.1
United Kingdom													
1969/71			10				818	225	937	574	2554	13008	881
1979/81			274				1058	262	949	751	3019	15917	838
1984/6			922				1066	294	982	876	3218	16160	752
1987			1353				1075	303	1010	994	3382	15358	793
1988			1039				960	325	1023	1079	3387	14981	795
1971–86 %			35.0				1.4	1.8		2.5	1.4	1.4	−1.1
United States													
1969/71	31174	121		1	3742	2225	9986	254	6225	4559	21023	53173	4045
1979/81	54861	2347			5034	3004	9991	147	7234	6712	24084	58139	4121
1984/6	53521	1448		1	4303	2623	10951	169	6528	7897	25546	63802	4040
1987	52330	1183		1	5234	3214	10734	144	6487	9335	26700	64662	4104
1988	41876	736		1	5492	3363	10732	153	7054	9517	27455	66010	4046
1971–86 %	3.5	15.4					1.4	−2.8	0.9	4.1	1.4	1.5	
USSR													
1969/71	521	6055	4		4006	2132	5496	968	4638	1040	12142	82533	2302
1979/81	494	4903	17		5848	2733	6694	858	5188	2143	14882	90934	3829
1984/6	547	5020	80		4867	2591	7458	857	5925	2830	17070	99587	4371
1987	712	6075	296		4449	2460	8125	895	6200	3000	18220	103360	4605
1988	760	6200	425		4650	2700	8359	869	6300	3127	18655	106438	4744
1971–86 %		−1.4	18.9		0.9	0.7	1.7	−0.9	0.8	6.7	1.8	1.0	3.7

	C1	C2	C3	C4	C5	C6	C7	C8	C9	C10	C11	C12	C13
Yugoslavia													
1969/71	5	334	13	2	7	4	261	52	572	138	1023	2759	137
1979/81	64	385	75	3	1		345	60	799	277	1481	4512	218
1984/6	209	279	127	2	1		353	60	893	312	1617	4690	235
1987	237	486	88	5	1		352	58	830	331	1571	4748	246
1988	180	410	68	3	1		370	57	850	336	1613	4850	242
1971–86 %	29.3		20.5		−18.1		1.9	1.3	2.9	5.7	3.1	3.4	3.3
EC-12													
1969/71	2	212	863	1130	320	165	6146	695	8213	3164	18218	100882	4214
1979/81	31	879	1748	1301	327	173	7481	881	11262	4966	24590	120818	4854
1984/6	471	2743	3633	1318	447	240	7975	963	12132	5316	26386	124878	4885
1987	1803	4106	5978	1681	482	255	8048	992	12726	5644	27409	117440	4835
1988	1652	3948	5335	1069	650	340	7796	1033	12869	5868	27567	115058	4965
1971–86 %	34.0	16.7	10.6	1.3	1.6	2.2	1.8	2.3	2.7	3.2	2.5	1.6	1.0
Other W. Europe													
1969/71	16	740	277	83	782	486	1155	461	1552	444	3612	22719	655
1979/81	71	1067	526	123	933	579	1372	467	2096	867	4802	27140	928
1984/6	338	1142	671	129	1010	629	1434	565	2258	965	5222	27635	1059
1987	487	1658	588	100	966	600	1437	569	2172	1002	5179	27348	1076
1988	340	1641	638	159	1055	655	1430	569	2204	1014	5216	27017	1079
1971–86 %	25.0	1.9	5.1				1.8	1.1	2.5	4.9	2.5	1.4	3.1
Total W. Europe													
1969/71	18	951	1139	1214	1102	651	7301	1155	9765	3608	21830	123601	4869
1979/81	102	1946	2274	1424	1260	752	8853	1348	13358	5833	29392	147958	5782
1984/6	809	3886	4304	1447	1457	868	9409	1528	14390	6281	31608	152513	5944
1987	2290	5764	6566	1781	1447	855	9485	1560	14897	6645	32588	144788	5911
1988	1992	5589	5973	1228	1704	995	9226	1602	15073	6882	32783	142075	6043
1971–86 %	27.8	9.0	9.4	1.2	0.9	1.0	1.8	1.9	2.6	3.4	2.5	1.6	1.3
North America													
1969/71	31431	160	1517	1	3742	2225	10874	261	6828	4985	22947	61420	4372
1979/81	55513	2530	2581		5034	3004	11010	152	8056	7244	26462	65969	4452
1984/6	54484	1508	3565	1	4303	2623	12040	177	7483	8480	28179	71772	4365
1987	53600	1235	3847	1	5234	3214	11762	151	7451	9989	29353	72648	4428
1988	43029	796	4243	1	5492	3363	11763	160	8052	10187	30163	74160	4369
1971–86 %	3.6	14.2	7.0				1.8	−2.6	1.2	4.0	1.4	1.4	1.4

Table A.5 (cont., page 5) Production: major oilseeds, cotton, livestock products (thousand tonnes and growth rates % p.a.)

	Soybeans	Sunflower seed	Rapeseed	Olive oil	Cotton seed	Cotton lint	Beef	Mutton	Pigmeat	Poultry-meat	Total Meat	Milk	Eggs
									carcass weight, excl. offals				
W.Eur.+ N.Am.													
1969/71	31449	1111	2656	1214	4844	2876	18175	1416	16593	8593	44777	185021	9241
1979/81	55615	4476	4855	1424	6294	3756	19863	1500	21414	13077	55855	213927	10234
1984/6	55293	5394	7870	1448	5760	3492	21448	1705	21873	14761	59787	224285	10310
1987	55890	7000	10412	1781	6681	4069	21248	1711	22348	16634	61941	217436	10339
1988	45021	6385	10216	1229	7196	4358	20989	1762	23124	17070	62945	216235	10412
1971–86 %	3.7	10.6	8.1	1.2			0.9	1.2	2.1	3.7	2.0	1.5	0.9
Eastern Europe													
1969/71	110	1366	755		26	13	1736	231	3914	764	6645	33746	1315
1979/81	528	1788	949		10	5	2091	259	5945	1663	9958	39668	1810
1984/6	502	2053	1911		11	6	2214	286	6095	1692	10288	42372	1915
1987	734	2378	2036		14	7	2250	280	6325	1762	10617	41996	1979
1988	518	2260	2124		10	5	2211	289	6164	1833	10497	41915	1869
1971–86 %	6.0	3.5	4.0		−5.8	−5.8	1.3	1.4	2.0	4.7	2.2	1.2	2.2
E.Europe + USSR													
1969/71	631	7421	759		4032	2145	7232	1199	8552	1804	18787	116279	3617
1979/81	1022	6691	966		5858	2738	8784	1117	11134	3806	24840	130602	5639
1984/6	1049	7073	1990		4878	2596	9671	1144	12020	4522	27357	141958	6286
1987	1446	8453	2332		4463	2467	10375	1175	12525	4762	28837	145356	6584
1988	1278	8460	2549		4660	2705	10570	1158	12464	4960	29152	148353	6613
1971–86 %	3.6		4.3		0.9	0.7	1.6	−0.4	1.4	5.9	2.0	1.1	3.3
All Study countries													
1969/71	32081	8532	3415	1214	8876	5021	25407	2615	25145	10397	63564	301300	12859
1979/81	56637	11167	5820	1424	12151	6493	28648	2617	32548	16883	80694	344528	15873
1984/6	56341	12467	9860	1448	10638	6088	31120	2849	33893	19283	87143	366243	16595
1987	57337	15452	12744	1781	11144	6536	31623	2886	34873	21396	90777	362792	16923
1988	46299	14845	12764	1229	11856	7063	31559	2920	35588	22029	92096	364588	17025
1971–86 %	3.7	2.8	7.2	1.2	0.9	1.1	1.1	0.6	1.8	4.2	2.0	1.3	1.7

All OECD countries													
1969/71	31578	782	2708	1206	4812	2857	19525	2652	16927	8979	48083	199747	10997
1979/81	55828	4206	4808	1414	6267	3744	22097	2607	22265	14068	61037	226872	12171
1984/6	55422	5295	7813	1440	5934	3618	23421	2914	22740	15975	65050	239571	12364
1987	56030	6614	10415	1772	6904	4220	23467	2957	23373	18013	67811	232118	12596
1988	45226	6121	10234	1220	7493	4556	23284	3005	24128	18484	68900	231628	12704
1971–86 %	3.7	12.2	7.9	1.3	1.9	1.7	0.9	0.8	2.2	3.9	2.0	1.4	0.9
World													
1969/71	43757	9926	6649	1426	22599	12223	38210	6939	36104	14741	95987	395211	20050
1979/81	85978	14399	11164	1667	27253	14358	44946	7577	51599	26467	130582	466375	27152
1984/6	95382	18777	18401	1700	31466	16892	47919	8482	59101	31966	147462	511560	32158
1987	100161	20582	22516	2000	30846	16533	49501	8861	62605	35652	156613	517968	34299
1988	92329	20947	21757	1462	33004	18031	49562	9000	64328	36795	159678	524717	35423
1971–86 %	5.2	4.7	7.3	1.0	1.9	1.7	1.4	1.6	3.3	5.4	2.9	1.8	3.2

Table A.6 Net trade balances (thousand tonnes) and self-sufficiency ratios (%)

	Wheat	Coarse grains	Total cereals	Sugar and products (raw sugar equiv.)	Oilseeds, veg. oils and products (oil equiv.)	Oilcakes and meals (product weight)	Meat and products (carcass weight, excl. offals)	Milk and products excl. butter (fresh milk equiv.)	Fruit and products (fresh fruit equiv.)	Coffee, cocoa and products (beans equiv.), tea
Albania										
Net trade 1969/71	-61	-5	-67	-16	-5	-5	-4	-8	14	-38
1984/6		4	2	-20	-4	-6	-1	-1	32	-80
1987		-1	-2	-25	-8	-6	-1		26	-84
1988	-50		-52	-45					17	-87
SSR (%) 1969/71	79	93	86	53	100	100	100	100		
1984/6	100	106	102	71	84	87	95	98		
Austria										
Net trade 1969/71	-37	-182	-268	-1	-81	-144	17	249	-452	
1984/6	564	263	770	64	-111	-485	82	638	-577	
1987	459	319	698	30	-102	-505	69	490	-701	
1988	808	331	1074	47	-96	-503	66	133	-681	
SSR (%) 1969/71	92	97	94	95	5	6	103	108		
1984/6	165	107	117	106	8	3	110	120		
Belgium-Lux.										
Net trade 1969/71	-839	-1815	-2689	218	-169	-555	190	349	-634	-86
1984/6	-479	-1095	-1674	629	-274	-129	251	426	-894	-96
1987	-427	-1717	-2202	646	-397	-162	309	1154	-1019	-108
1988	-777	-1447	-2292	177	-438	-307	383	698	-1119	-120
SSR (%) 1969/71	49	38	42	168	2	35	127	109		
1984/6	73	47	57	237	2	92	128	112		

	C1	C2	C3	C4	C5	C6	C7	C8	C9	C10
Bulgaria										
Net trade 1969/71	58	176	230	−274	67	−101	77	131	781	−8
1984/6	134	−810	−687	−401	−2	−400	150	51	764	−20
1987	−137	−583	−755	−435	−24	−307	135	95	632	−26
1988	−113	−763	−889	−371	−83	−471	133	50	607	−25
SSR (%) 1969/71	101	105	103	24	133	71	126	107		
1984/6	109	76	91	8	98	43	125	102		
Canada										
Net trade 1969/71	10833	2540	13314	−925	332	−83	−19	1523	−1481	−139
1984/6	18473	4526	22884	−1041	986	−282	241	963	−2673	−168
1987	22615	6228	28713	−834	1249	−116	185	426	−3015	−171
1988	20302	3413	23569	−829	1180	−163	276	577	−2618	−177
SSR (%) 1969/71	306	113	151	19	235	89	99	119		
1984/6	456	126	199	10	233	85	110	112		
Cyprus										
Net trade 1979/81	−59	−237	−299	−17	−10	−31	−7	−27	179	−2
1984/6	−82	−343	−427	−21	−12	−48	−6	−22	198	−2
1987	−87	−363	−453	−24	−12	−56	−8	−21	178	−3
1988	−86	−346	−435	−20	−11	−52	−7	−30	156	−3
SSR (%) 1979/81	27	29	28		35		86	80		
1984/6	16	22	21		30		89	85		
Czechoslovakia										
Net trade 1969/71	−1159	−188	−1424	97	−121	−408	−91	36	−395	−40
1984/6	−60	95	−33	107	−72	−548	23	917	−332	−52
1987	−63	152	31	77	−72	−477	28	924	−420	−56
1988	−257	177	−140	−38	−57	−482	47	925	−425	−62
SSR (%) 1969/71	76	99	87	114	24	28	92	101		
1984/6	95	109	101	137	67	33	101	115		
Denmark										
Net trade 1969/71	48	−406	−360	51	−74	−596	777	742	−233	−73
1984/6	379	950	1330	292	63	−1967	954	2500	−415	−69
1987	488	948	1434	296	12	−2230	944	2450	−523	−67
1988	676	1354	2028	266	1	−2052	959	2242	−524	−70

Table A.6 (cont., page 2) Net trade balances (thousand tonnes) and self-sufficiency ratios (%)

		Wheat	Coarse grains	Total cereals	Sugar and products (raw sugar equiv.)	Oilseeds, veg. oils and products (oil equiv.)	Oilcakes and meals (product weight)	Meat and products (carcass weight, excl. offals)	Milk and products excl. butter (fresh milk equiv.)	Fruit and products (fresh fruit equiv.)	Coffee, cocoa and products (beans equiv.), tea
SSR (%)	1969/71	120	93	94	119	16	42	357	120		
	1984/6	121	126	125	254	153	7	333	189		
Finland											
Net trade	1969/71	49	25	58	−160	−29	9	26	299	−212	−55
	1984/6	−30	681	634	−54	−25	−1	38	618	−320	−64
	1987		237	215	−98	−20		37	592	−411	−75
	1988	−112	−65	−201	−67	−32	−5	17	416	−396	−56
SSR (%)	1969/71	114	106	106	29	12	112	113	111		
	1984/6	94	119	114	66	61	99	113	123		
France											
Net trade	1969/71	4718	6140	10776	581	−533	−1193	−125	4459	−1758	−313
	1984/6	16033	10426	26202	2185	46	−3637	192	4586	−980	−413
	1987	15834	11706	27291	2070	798	−3869	402	3891	−573	−461
	1988	16336	12327	28395	2262	1035	−3685	476	3561	−1145	−476
SSR (%)	1969/71	147	150	148	136	36	40	96	119		
	1984/6	237	179	204	176	103	25	103	116		
German DR											
Net trade	1969/71	−1697	−796	−2534	−162	−137	−546	−7	−47	−688	−62
	1984/6	−929	−1578	−2543	−58	−85	−881	106	−3	−676	−100
	1987	−447	−646	−1120	−29	−53	−829	172	−1	−702	−111
	1988	−508	−1524	−2060	−128	−73	−493	157	−9	−688	−109
SSR (%)	1969/71	60	86	75	76	34	28	102	99		
	1984/6	89	80	82	102	60	23	105	100		

	1	2	3	4	5	6	7	8	9	10
Germany, FR										
Net trade 1969/71	−1096	−3960	−5157	−91	−1022	−2403	−570	331	−5152	−470
1984/6	−214	−1455	−1794	864	−1241	−3256	−474	4970	−5244	−628
1987	888	−588	169	1295	−1232	−2631	−426	8558	−6165	−678
1988	2046	−268	1655	1060	−1251	−2925	−569	10034	−5889	−673
SSR (%) 1969/71	84	76	78	95	7	44	88	103		
1984/6	104	92	96	142	20	53	91	126		
Greece										
Net trade 1969/71	18	−280	−256	−28	−13	−1	−131	−206	1048	−19
1984/6	644	−37	621	−8	125	18	−225	−701	1650	−37
1987	172	−279	−76	−45	83	−16	−289	−739	1377	−42
1988	671	−88	617	−164	−4	19	−303	−789	1452	−43
SSR (%) 1969/71	107	82	96	86	94	101	69	87		
1984/6	123	100	109	89	135	104	69	71		
Hungary										
Net trade 1969/71	75	−35	21	−61	22	−321	163	−271	566	−34
1984/6	1684	232	1890	24	205	−653	518	101	923	−65
1987	1281	−425	859	27	278	−620	611	50	870	−72
1988	1828	60	1899	95	304	−642	620	103	901	−72
SSR (%) 1969/71	104	105	104	74	166	22	120	87		
1984/6	135	103	115	100	273	37	143	104		
Iceland										
Net trade 1969/71	−14	−26	−41	−11	−1	−1	4	20	−11	−3
1984/6	−17	−33	−52	−13	−2	−1	2	14	−18	−4
1987	−21	−47	−70	−13	−3		4	6	−21	−4
1988	−24	−46	−72	−13	−3	−2	2	6	−28	−4
SSR (%) 1969/71							129	121		
1984/6							112	113		
Ireland										
Net trade 1969/71	−128	−293	−426	−18	−22	−119	330	661	−135	−18
1984/6	−336	139	−220	44	−50	−428	428	3831	−252	−22
1987	−304	343	13	35	−63	−441	505	3502	−265	−17
1988	−322	445	92	57	−62	−503	438	3056	−252	−22

Table A.6 (cont., page 3) Net trade balances (thousand tonnes) and self-sufficiency ratios (%)

		Wheat	Coarse grains	Total cereals	Sugar and products (raw sugar equiv.)	Oilseeds, veg. oils and products (oil equiv.)	Oilcakes and meals (product weight)	Meat and products (carcass weight, excl. offals)	Milk and products excl. butter (fresh milk equiv.)	Fruit and products (fresh fruit equiv.)	Coffee, cocoa and products (beans equiv.), tea
SSR (%)	1969/71	72	80	77	82		5	251	120		
	1984/6	63	113	95	114	6	1	228	238		
Israel											
Net trade	1969/71	−355	−766	−1150	−148	−56	45	−35	−151	1125	−12
	1984/6	−643	−1127	−1824	−336	−82	2	−15	−83	1247	−25
	1987	−576	−1391	−2029	−269	−101	−7	−7	−69	1367	−24
	1988	−594	−1267	−1922	−231	−95	−17	6	−94	1292	−39
SSR (%)	1969/71	31	5	15	8	27	130	76	81		
	1984/6	21	3	9		30	102	93	92		
Italy											
Net trade	1969/71	−634	−5697	−6029	−363	−668	−233	−701	−3121	2440	−201
	1984/6	−2500	−2064	−4160	−283	−656	−1399	−1108	−7290	4021	−320
	1987	−2654	−2376	−4604	−3	−586	−1761	−1542	−6817	3231	−328
	1988	−2081	−2362	−4059	106	−528	−1491	−1493	−6871	3296	−322
SSR (%)	1969/71	92	49	72	77	48	83	75	76		
	1984/6	81	80	83	88	47	56	75	62		
Malta											
Net trade	1969/71	−57	−56	−114	−16	−4	−16	−8	−35	−10	−2
	1984/6	−33	−84	−118	−18	−5	−7	−9	−52	−18	−2
	1987	−46	−91	−139	−23	−6	−11	−10	−55	−19	−1
	1988	−59	−105	−166	−22	−6	−10	−9	−61	−25	−2
SSR (%)	1969/71	4	3	3				48	44		
	1984/6	10	4	6				56	37		

	C1	C2	C3	C4	C5	C6	C7	C8	C9	C10
Netherlands										
Net trade 1969/71	-769	-2126	-2936	44	-417	-552	631	1485	-587	-121
1984/6	-809	-2511	-3385	151	-660	-1792	1524	1789	-1406	-159
1987	-876	-2176	-3118	598	-914	-2090	1792	1284	-1539	-171
1988	-1264	-2215	-3560	415	-888	-2322	1805	2915	-1728	-204
SSR (%) 1969/71	48	30	35	99	5	66	184	129		
1984/6	54	13	30	163	4	62	238	125		
Norway										
Net trade 1969/71	-374	-288	-670	-182	-63	-68	-2	124	-217	-44
1984/6	-263	-159	-433	-181	-71	51	-2	197	-337	-55
1987	-275	-291	-580	-183	-75	65	-1	237	-397	-55
1988	-306	-290	-611	-170	-69	12	7	224	-373	-51
SSR (%) 1969/71	3	71	53		3	72	99	108		
1984/6	39	97	81		7	124	99	110		
Poland										
Net trade 1969/71	-1399	-884	-2349	230	-60	-291	106	310	-134	-61
1984/6	-1896	-79	-2052	189	21	-948	59	1443	349	-89
1987	-2343	-592	-3009	169	72	-1320	191	1399	240	-90
1988	-2316	-529	-2909	39	18	-1295	169	1295	204	-102
SSR (%) 1969/71	79	93	89	114	74	54	106	102		
1984/6	80	97	91	106	105	34	104	110		
Portugal										
Net trade 1969/71	-314	-510	-838	-213	-73	-56	-18	16	268	-21
1984/6	-667	-1698	-2439	-287	-133	141	-22	-114	195	-27
1987	-549	-740	-1338	-264	-182	-43	-39	-40	168	-35
1988	-438	-857	-1385	-314	-231	-69	-56	2	128	-37
SSR (%) 1969/71	73	66	70	5	49	67	94	102		
1984/6	41	34	37	2	31	121	95	89		
Romania										
Net trade 1969/71	120	501	576	-33	141	-42	81	-10	413	-12
1984/6	142	-344	-238	-119	-12	-80	193	36	341	-22
1987	30	368	322	-160	-14	-44	220	25	413	-17
1988	250	64	265	-148	-31	-69	164	-129	397	-12

Table A.6 (cont., page 4) Net trade balances (thousand tonnes) and self-sufficiency ratios (%)

		Wheat	Coarse grains	Total cereals	Sugar and products (raw sugar equiv.)	Oilseeds, veg. oils and products (oil equiv.)	Oilcakes and meals (product weight)	Meat and products (carcass weight, excl. offals)	Milk and products excl. butter (fresh milk equiv.)	Fruit and products (fresh fruit equiv.)	Coffee, cocoa and products (beans equiv.), tea
SSR (%)	1969/71	101	103	102	92	169	90	113	100		
	1984/6	102	99	100	86	98	88	111	101		
Spain											
Net trade	1969/71	378	−2429	−2000	−161	−31	−95	−97	−831	2250	−106
	1984/6	−210	−2634	−2816	53	44	−671	−105	−1186	4017	−183
	1987	−76	−420	−416	57	90	−970	−224	−520	3993	−217
	1988	−413	−547	−903	1	176	−1604	−188	−566	3863	−213
SSR (%)	1969/71	105	73	84	88	82	93	93	86		
	1984/6	98	83	87	98	106	77	96	86		
Sweden											
Net trade	1969/71	221	411	611	−118	−48	−312	31	−195	−589	−120
	1984/6	624	601	1187	−36	−36	−184	69	197	−670	−110
	1987	682	283	925	−37	−57	−135	8	177	−797	−113
	1988	431	133	522	−49	−73	−256	3	17	−811	−109
SSR (%)	1969/71	122	115	116	63	79	22	108	94		
	1984/6	169	119	128	97	74	47	113	106		
Switzerland											
Net trade	1969/71	−478	−998	−1499	−234	−87	−89	−74	264	−671	−47
	1984/6	−301	−836	−1198	−172	−82	−42	−57	505	−752	−47
	1987	−314	−716	−1092	−140	−81	−37	−60	456	−812	−60
	1988	−257	−710	−998	−135	−75	−55	−79	380	−823	−54
SSR (%)	1969/71	45	24	32	20	9	44	83	108		
	1984/6	66	37	47	47	17	73	89	114		

		(1)	(2)	(3)	(4)	(5)	(6)	(7)	(8)	(9)	(10)
Turkey											
Net trade	1969/71	-754	7	-758	77	-2	272	14	-80	549	7
	1984/6	-256	-166	-515	298	-225	8	64	-70	968	-7
	1987	178	7	10	-159	-248	-63	27	-69	993	-14
	1988	2542	236	2677	45	-374	-4	46	-39	1253	-11
SSR (%)	1969/71	99	103	100	105	97	228	102	98		
	1984/6	93	101	96	105	74	101	107	99		
United Kingdom											
Net trade	1969/71	-4803	-4235	-9145	-1806	-753	-901	-1430	-1940	-3167	-423
	1984/6	1300	2086	3208	-952	-706	-2022	-730	908	-4300	-421
	1987	2482	1570	3881	-858	-902	-2081	-617	250	-4761	-407
	1988	379	1270	1468	-946	-763	-1959	-716	-243	-4888	-414
SSR (%)	1969/71	48	72	62	36	1	29	64	87		
	1984/6	132	121	126	61	35	27	83	104		
United States											
Net trade	1969/71	16903	17855	36260	-4647	2148	4339	-936	-288	-931	-1728
	1984/6	31891	47158	81022	-2423	4054	5721	-1027	-46	-8443	-1720
	1987	32008	48388	82647	-863	3855	7483	-1115	-1074	-10435	-1842
	1988	41594	54373	97969	-1015	3312	7829	-1046	-589	-7148	-1518
SSR (%)	1969/71	187	114	124	52	142	133	96	99		
	1984/6	216	149	159	74	162	130	96	97		
USSR											
Net trade	1969/71	5308	430	5412	-820	488	52	-121	-40	-2339	-179
	1984/6	-19093	-15989	-35380	-4931	-970	-366	-1002	-481	-2802	-327
	1987	-16577	-11686	-28810	-4897	-1425	-3013	-1022	-454	-1771	-363
	1988	-19763	-13464	-33720	-4015	-776	-3277	-869	-342	-1727	-334
SSR (%)	1969/71	100	100	100	83	117	101	102	100		
	1984/6	83	85	84	64	77	93	95	100		
Yugoslavia											
Net trade	1969/71	-199	-89	-321	-32	-41	-143	130	-104	30	-49
	1984/6	-76	1110	1022	-39	-179	-116	86	-126	387	-45
	1987	-514	912	394		-47	-57	52	-132	328	-67
	1988	415	-261	150	74	7	-74	-11	-194	343	-70

Table A.6 (cont., page 5) Net trade balances (thousand tonnes) and self-sufficiency ratios (%)

	Wheat	Coarse grains	Total cereals	Sugar and products (raw sugar equiv.)	Oilseeds, veg. oils and products (oil equiv.)	Oilcakes and meals (product weight)	Meat and products (carcass weight, excl. offals)	Milk and products excl. butter (fresh milk equiv.)	Fruit and products (fresh fruit equiv.)	Coffee, cocoa and products (beans equiv.), tea
SSR (%) 1969/71	97	104	101	75	74	58	116	96		
1984/6	89	99	96	105	50	82	105	97		
EC–12										
Net trade 1969/71	−3421	−15610	−19060	−1786	−3776	−6705	−1144	1944	−5658	−1849
1984/6	13141	2107	14873	2688	−3442	−15142	684	9718	−3606	−2374
1987	14980	6271	21035	3825	−3293	−16293	815	12973	−6074	−2532
1988	14814	7611	22057	2919	−2953	−16897	736	14038	−6805	−2594
SSR (%) 1969/71	92	82	86	87	32	52	94	103		
1984/6	129	104	114	124	55	50	103	109		
Other W. Europe										
Net trade 1969/71	−2098	−2049	−4341	−859	−420	−445	97	363	−203	−364
1984/6	−512	−89	−952	−528	−834	−779	249	1807	221	−439
1987	−514	−1140	−2123	−941	−754	−815	110	1612	−266	−500
1988	2710	−2391	−32	−586	−835	−971	39	757	−75	−486
SSR (%) 1969/71	93	95	94	63	63	75	103	102		
1984/6	94	97	95	84	56	76	104	107		
Western Europe										
Net trade 1969/71	−5519	−17659	−23400	−2644	−4196	−7150	−1047	2307	−5861	−2213
1984/6	12629	2018	13922	2160	−4276	−15921	933	11525	−3386	−2813
1987	14466	5131	18911	2884	−4047	−17108	925	14586	−6341	−3032
1988	17524	5220	22024	2334	−3788	−17868	775	14796	−6881	−3080
SSR (%) 1969/71	93	85	88	83	37	55	95	102		
1984/6	117	102	108	114	55	53	103	109		

North America										
Net trade 1969/71	27736	20395	49573	-5572	2480	4256	-954	1235	-2411	-1867
1984/6	50365	51684	103906	-3464	5040	5439	-785	917	-11116	-1888
1987	54623	54616	111360	-1697	5104	7367	-930	-648	-13449	-2012
1988	61896	57786	121538	-1844	4493	7665	-771	-11	-9766	-1695
SSR (%) 1969/71	208	114	127	48	149	130	96	101		
1984/6	254	146	164	65	170	126	97	99		
W.Eur.+ N.Am.										
Net trade 1969/71	22217	2736	26171	-8216	-1716	-2893	-2001	3542	-8272	-4079
1984/6	62993	53701	117824	-1303	764	-10482	147	12441	-14502	-4700
1987	69089	59747	130268	1187	1057	-9741	-6	13938	-19790	-5044
1988	79419	63004	143559	490	704	-10203	4	14784	-16647	-4775
SSR (%) 1969/71	124	102	108	68	87	90	96	102		
1984/6	157	129	137	97	107	81	100	105		
Eastern Europe										
Net trade 1969/71	-4003	-1227	-5480	-202	-88	-1708	328	150	542	-218
1984/6	-924	-2484	-3663	-258	54	-3511	1050	2544	1368	-347
1987	-1679	-1725	-3673	-350	186	-3598	1357	2492	1032	-372
1988	-1116	-2514	-3834	-552	78	-3451	1290	2235	997	-381
SSR (%) 1969/71	85	97	93	94	90	45	107	100		
1984/6	99	95	96	98	104	42	111	106		
E.Europe + USSR										
Net trade 1969/71	1305	-796	-68	-1022	400	-1657	207	110	-1797	-396
1984/6	-20017	-18473	-39043	-5189	-916	-3877	48	2063	-1434	-674
1987	-18256	-13410	-32482	-5247	-1239	-6610	335	2038	-739	-735
1988	-20879	-15978	-37554	-4566	-698	-6728	422	1893	-730	-716
SSR (%) 1969/71	97	99	98	86	110	77	104	100		
1984/6	87	89	88	73	84	65	100	102		
All Study countries										
Net trade 1969/71	23522	1940	26103	-9238	-1316	-4550	-1793	3652	-10070	-4476
1984/6	42976	35227	78782	-6492	-153	-14359	195	14504	-15935	-5374
1987	50833	46336	97785	-4061	-182	-16351	329	15976	-20528	-5779
1988	58540	47026	106004	-4077	7	-16931	426	16677	-17377	-5490

Table A.6 (cont., page 6) Net trade balances (thousand tonnes) and self-sufficiency ratios (%)

		Wheat	Coarse grains	Total cereals	Sugar and products (raw sugar equiv.)	Oilseeds, veg. oils and products (oil equiv.)	Oilcakes and meals (product weight)	Meat and products (carcass weight, excl. offals)	Milk and products excl. butter (fresh milk equiv.)	Fruit and products (fresh fruit equiv.)	Coffee, cocoa and products (beans equiv.), tea
SSR (%)	1969/71	109	101	104	74	93	88	98	101		
	1984/6	121	114	117	87	102	78	100	104		
All OECD countries											
Net trade	1969/71	25730	-5075	22557	-8990	-2752	-2779	-997	9672	-10133	-4228
	1984/6	72500	38733	112687	-381	-1013	-10357	1116	18397	-17715	-5116
	1987	80015	41486	123164	1984	-1067	-10092	1184	19679	-23192	-5515
	1988	86631	44498	132616	1262	-1552	-10938	1055	20335	-20048	-5255
SSR (%)	1969/71	126	100	107	70	81	91	98	105		
	1984/6	163	124	134	100	98	81	102	108		

Notes to tables

No entry in a table means either zero or a value less than one half the unit employed. For growth rates, no entry means a growth rate which is actually zero or statistically not significantly different from zero, at 5% level of significance.

...   Data not available.

Table A.1

Population from 1985 from the UN Population Division supplemented with more recent national estimates (for explanations see FOA, *Production Yearbook 1987*, Rome, 1988) 2000 projections from UN, *World Population Prospects: Estimates and projections as assessed in 1984*, Population Studies No. 98, New York, 1986 (medium variant projections). GNP from World Bank, *World Development Report 1988*, Oxford University Press, London, 1988.

Table A.2

Arable land refers to land under temporary crops (double-cropped areas are counted only once), temporary meadows for mowing or pasture, land under market and kitchen gardens (including cultivation under glass), and land temporarily fallow or lying idle.
Land under permanent crops refers to land cultivated with crops that occupy the land for long periods and do not need to be replanted after each harvest; it includes land under shrubs, fruit trees, nut trees and vines, but excludes land under trees grown for wood or timber.

Table A.4

Production data shown come from the latest (May 1989) revision of the general FAO data set. As such there may be small differences from the data used in the analysis and occasionally referred to in the text.

Table A.6

The net trade data for 1987 and 1988 (but not the earlier three-year averages for 1969/71 and 1984/6) come likewise from the latest (September 1989) revision of the general FAO data set.

SSR

Self-Sufficiency Ratio, i.e. production percent of total domestic demand. The latter is equal to the left hand side of the Supply Utilization Account equation given in Appendix 2.

REFERENCES

Aganbeguian, A. 1987, *Perestroïka: Le double défi sovietique*, Economica, Paris.

Aganbegyan, A. 1988, 'The Economics of Perestroika', *International Affairs*, Spring.

Aganbegyan, A. 1989, *Economic Restructuring in the USSR and International Economic Relations*, Paper presented to the 9th World Congress of the International Economic Association, Athens, 28 August–1 September 1989.

Alexandratos, N. 1988 (ed.), *World Agriculture: Toward 2000, an FAO Study*, Belhaven Press, London, and New York University Press, New York. French edition: Alexandratos, N. (sous la direction de) 1989, *Agriculture mondiale: Horizon 2000, étude de la FAO*, Economica, Paris.

Ambassade de l'URSS, Berne 1989, 'Rapport de Michail Gorbatchev au Plenum de CC du PCUS', *Les Nouvelles, Bulletin édité par l'Ambassade de l'URSS*, No. 3230, 17 March 1989, Berne.

Anderson, K. and Y. Hayami 1986, *The Political Economy of Agricultural Protection, East Asia in International Perspective*, Allen and Unwin, London.

Barse, J., W. Ferguson and R. Seem 1988, *Economic Effects of Banning Soil Fumigants*, USDA, Agricultural Economic Report 602, Washington DC

Blandford, D., de Gorter H. and D. Harvey 1989 'Farm Income Support with Minimal Trade Distortions', *Food Policy*, August.

Blaylock, J. and D. Smallwood 1986, *US Demand for Food: Households expenditures, demographics, and projections*, USDA, Economic Research Service, Technical Bulletin, No. 1713, Washington, DC

Bloom, B. 1989, 'The Great British Food Debate', *Financial Times*, 28 February 1989.

Borsody, L. 1987, 'Forecasting USSR Grain Imports', *Food Policy*, May.

Boulatov, D. 1989 'Pourquoi nous avons besoin de la FAO', *Temps Nouveaux*, No. 35–89, Moscow.

Chernyak, A. 1988, 'Mouths to Feed by Statistics and in Real Life: The Food Issue through the Prism of Social Justice', *Pravda*, 1 September 1988.

Clark, B. and T. Dickson 1989, 'Towards a Green Consensus', *Financial Times*, 12 April 1989.

Coffman, R. 1983, 'Plant Research and Technology', in Rosenblum, J. (ed.), *Agriculture in the Twenty-First Century*, Wiley, New York.

Commins, P. and J. Higgins 1987, *Farming and Farm-Workers in the European Community 1985–2000*, Commission of the European Communities, FAST Occasional Papers, No. 152, Brussels.

Commission of the European Communities (CEC) 1980, 'Effects on the Environment of the Abandonment of Agricultural Land', *Information on Agriculture* 62, Brussels.

Commission of the European Communities (CEC) 1984, *Public Expenditure on Agriculture*, Brussels, November 1984.

Commission of the European Communities (CEC) 1987, *Study on Bioethanol, Final Report*. Report prepared for the EC Commission by Agro Developpement *et al.*, June 1987. CM22/87, Brussels.

Commission of the European Communities (CEC) 1988a, *The Agricultural Situation in the Community—1987 Report*, Brussels.

Commission of the European Communities (CEC) 1988b, 'The Future of the European Food System', Executive summary of the ALIM subprogramme of the *FAST Programme II* (1984–7), Brussels.

Commission of the European Communities (CEC) 1988c, *Commission Proposals on the Prices for Agricultural Products and on Related Measures (1988/89)*, COM (88) 120 final-I, Brussels.

Commission of the European Communities (CEC) 1988d, 'The Economics of 1992', Special issue of the *European Economy*, No. 35 (March).

Commission of the European Communities (CEC) 1988e, *Disharmonies in EC and US Agricultural Policy Measures* (2 vols), Brussels.

Commission of the European Communities (CEC) 1988f, 'The Future of Rural Society, Commission Communication to Parliament and the Council', *Bulletin of the European Communities*, Supplement 4/88.

Commission of the European Communities (CEC) 1989a, *The Agricultural Situation in the Community, 1988 Report*, Brussels.

Commission of the European Communities (CEC) 1989b, Court of Auditors, 'Special Report No. 1/89 on The Agrimonetary System Accompanied by the Replies of the Commission', *Official Journal of the European Communities*, C 128, Vol. 32, 24 May 1989.

Commission of the European Communities (CEC) 1989c, *Adjustment of the Agricultural Structures Policy*, COM (89) final, 3 July 1989.

Commission of the European Communities (CEC) 1989d, *The Impact of Biotechnology on Agriculture in the EC to the Year 2005*, Study prepared by the Bureau Européen de Recherches SA, Brussels.

Cook, E. 1987, 'Soviet Food Markets: Will the Situation Improve Under Gorbachev?', *Comparative Economic Studies*, 29 (1).

Crocker, T. 1989, 'The Impact of Pollution from other Sources on Agriculture', in OECD 1989b.

Crosson, P. 1989, 'Dryland Farming, Soil Conservation and Erosion', in OECD 1989b.

Csáki, C. 1987, *Agricultural Situation in Eastern Europe: Supply-demand trends and reforms in the late 1980s*, Paper prepared for the 'Conference on Economic Dilemmas of Eastern Europe in the Late 1980s: Reform, Trade and Finance', Woodrow Wilson Center, Washington DC, September.

Daberkow, S. and K. Reichelderfer 1988, 'Low-Input Agriculture: Trends, Goals and Prospects for Input Use', *American Journal of Agricultural Economics*, December.

Davenport, M. 1988, *European Community Trade Barriers to Tropical Agricultural Products*, Overseas Development Institute, Working Paper 27, London.

De Haen, H. 1989, 'Intensive Crop Production and the Use of Agricultural Chemicals', in OECD 1989b.

Desai, P. 1986, *Weather and Grain Yields in the Soviet Union*, IFPRI Research Report No. 54, Washington, DC.

Diamond, D., L. Bettis and R. Ramson 1983, 'Agricultural Production', in Bergson, B. and H. Levine (eds), *The Soviet Economy Toward the Year 2000*, Allen and Unwin, London.

Dicks, M., F. Llacuna and M. Linsenbigler (1988), *The Conservation Reserve Program: Implementation and Accomplishments, 1986–1987*, USDA, ERS, Statistical Bulletin No. 763, Washington, DC.

Douw, L., L.B. van der Giessen and J.H. Post 1987, *De Nederlandse Landbouw na 2000: Een verkenning*, Landbouw-Economisch Instituut, Mededeling 379, The Hague.

Dubgaard, A. 1989, 'Structural Change in the Danish Agro-Industrial Complex, 1966–1987', in EAAE (1989).

Dunne, N. 1989, 'Subsidies Have Failed to Save the Family Farm', *Financial Times*, 9 August 1989.

EAAE 1989, *Integration and Cooperation in the Agro-Food Industry*, Proceedings of the 21st Seminar of the European Association of Agricultural Economists, Kiev, 3–7 October 1989, Institute of Economics, Academy of Sciences of the Ukrainian SSR, Kiev.

Ennew, C.T. 1987, 'A Model of Import Demand for Grain in the Soviet Union', *Food Policy*, May.

Ervin, D. and M. Dicks 1988, 'Cropland Diversion for Conservation and Environmental Improvement: An Economic Welfare Analysis', *Land Economics*, Vol. 64, No. 3.

EUROSTAT 1986, *National Accounts ESA, Input–Output Tables 1980*, Luxembourg.

Evenson, R. 1986, 'The Economics of Extension', in Jones, G. (ed.), *Investing in Rural Extension: Strategies and Goals*, Elsevier, London.

FAO 1973, *Agricultural Protection: Domestic policy and International Trade* Rome (Document C 73/LIM/9).

FAO 1987, *Agricultural Price Policies: Issues and Proposals*, Rome.

FAO 1988, *The Dynamics of Agrarian Structures in Europe: Case Studies of Germany, FR, Hungary, Italy, Norway and Poland*, Rome.

FAO 1989, *Interpretation of the International Undertaking on Plant Genetic Resources* (Document C 89/24), Rome.

FAO/CCP 1988, *Utilization of Grains as Animal Feed: Recent Trends and Issues*, Rome (Document CCP: GR 88/5).

FAO/CCP 1989, *Selected Issues in Agricultural Policy Reform: Strengthening market orientation of agricultural policies and the possible role of direct support*, Rome (Document CCP 89/19).

FAO/CFS 1987, *Impact on World Food Security of Agricultural Policies in Industrialized Countries*, Rome (Document CFS 87/3).

FAO/ERC 1988, *Integration of Environmental Aspects in Agricultural, Forestry and Fisheries Policies in the Region*, Document ERC/88/3 of the Sixteenth FAO Regional Conference for Europe, Cracow, 23–26 August 1988.

FAO/ERC 1990, *Socio-economic Aspects of Environmental Measures Related to European Agriculture*, (Document ERC/90/3 – forthcoming), Rome.

FAPRI 1988, *Ten-Year International Agricultural Outlook, Summary and Tables*, Food and Agricultural Policy Research Institute, Iowa State University (CARD) and University of Missouri (CNFAP), March.

Fleisher, B. 1989, 'The Economic Risks of Deliberately Released Genetically Engineered Micro-Organisms', *American Journal of Agricultural Economics*, May.

Fowler, G., E. Lachkovics, P. Mooney and H. Shand 1988, 'The Laws of Life, Another Development and the New Biotechnologies', *Development Dialogue*, 1–2.

Gardner, B. 1988, *Domestic Policies to Make Trade Liberalization Politically Possible: U.S. Case*, Paper for the International Agricultural Trade Research Consortium, Symposium on Bringing Agriculture into the GATT, Annapolis, Maryland, 19–20 August 1988.

Gardner, B. 1989, 'Economic Implications of a Changing World for US Agriculture in the 1990s', *Annual Agricultural Outlook Conference*, USDA, Washington DC, November.

Gorbachev, M. 1988, *Perestroïka: New thinking for our country and the world*, Fontana Paperbacks, London.

Greenfield, J. 1989, *Modelling the Grain Imports of China and the USSR: Discussion Note*, FAO, Rome.

Guzhvin, P. 1987, 'The Yield of Land and the Structure of Investment', *Problems of Economics* (February; translation of 'Otdacha Zemli i Struktura Vlozhenii', *Kommunist*, No. 8, 1986).

Hartmann, M., W. Henrichsmeyer and P.M. Schnitz 1989, *Political Economy of the CAP in the European Community*, Paper presented to 9th World Congress of the International Economic Association, Athens, 28 August–1 September 1989.

Harvey, D. 1988, *Decoupling and the European CAP*, Paper for the Symposium of the International Agricultural Trade Consortium on Bringing Agriculture into the GATT, Annapolis, USA, 19–20 August 1988.

Harvey, D. and J. Hall 1989, *PSEs, Producer Benefits and Transfer Efficiency of the CAP and Alternatives*, Paper presented to the Seminar of the European Association of Agricultural Economists on Costs and Benefits of Agricultural Policies and Projects in Europe, Amsterdam, 12–13 October 1989.

Harvey, D. and M. Whitby 1988, 'Issues and Policies', in Whitby, M. and J. Ollerenshaw (eds), *Land Use and the European Environment*, Belhaven Press, London.

Hayami, Y. 1988, *Japanese Agriculture under Siege, The Political Economy of Agricultural Policies*, St. Martin's Press, New York.

Henrichsmeyer, W. and A. Ostermeyer-Schloeder, 1987, *Productivity Growth and Factor Adjustment in EC Agriculture*, Paper presented to the 5th Congress of the European Association of Agricultural Economists, Hungary, Balatonszéplak.

Hine, R., K. Ingersent and A. Rayner 1989, 'Agriculture in the Uruguay Round: From the Punta del Este to the Geneva Accord', *Journal of Agricultural Economics*, 40, 3, September.

HMSO 1988, *The Nitrate Issue*, Her Majesty's Stationery Office, London (abstracted in Environmental Data Service—ENDS, Report 167, December 1988).

Hooke, A. 1989, *Demand for Australia's Agricultural Exports in Developing Asia*, ACC/Westpac Economic Discussion Paper No. 1, Australian Chamber of Commerce and Westpac Banking Corporation, Canberra.

Houck, J. 1988, 'Link between Agricultural Assistance and International Trade', *Agricultural Economics*, 2 (1988).

IATRC 1988, *Designing Acceptable Agricultural Policies*, Summary Report prepared by the International Agricultural Trade Research Consortium, Symposium on Bringing Agriculture into the GATT, Annapolis, Maryland, 19–20 August 1988.

International Wheat Council (IWC) 1983, 'Long Term Grain Outlook', *Secretariat Paper*, No. 14, London.

International Wheat Council (IWC) 1987, 'Long Term Outlook for Grain Imports by Developing Countries', *Press Release*, November 1987, London.

International Wheat Council (IWC) 1988, 'Long Term Outlook for Grain Imports by Developing Countries', *Secretariat Paper*, No. 17, London.

International Wheat Council (IWC) 1989, *Market Report, PMR 180*, 28 November.

Jampel, W. and E. Lhomel 1989, 'L'Industrie Agro-alimentaire en Europe de l'Est, Production et Échanges', *Le Courrier des Pays de l'Est*, No. 338, March.

Johnson, D. Gale 1989, *Soviet and Chinese Agriculture and Trade*, Paper No. 89:14, Office of Agricultural Economics Research, University of Chicago (presented to the Joint Economic Committee of the US Congress, 7 September 1989).

Knutson, R., J. Penn and W. Boehm 1983, *Agricultural and Food Policy*. Prentice-Hall, Englewood Cliffs, NJ.

Köhler, G. 1989, 'Economic and Social Development Aims in the German Democratic Republic', in UN/ECE, *Economic Reforms in the European Centrally Planned Economies*, Economic Studies, No. 1, United Nations, New York.

Konandreas, P. and R. Perkins 1989, *Some Implications of Trade Liberalization in Cereals for Low Income Food-deficit Countries*, Paper for the Joint OECD/World Bank Symposium on Agricultural Trade Liberalization: Implications for Developing Countries, OECD, Paris, 5–6 October 1989.

Kornai, J. 1986, 'The Hungarian Reform Process: Visions, Hopes, and Reality', *Journal of Economic Literature*, (December)

Krueger, A., M. Schiff and A. Valdés 1988, 'Agricultural Incentives in Developing Countries: Measuring the Effect of Sectoral and Economy-wide Policies', *World Bank Economic Review*, 2, 3, September.

Kuba, F. 1988, 'Agricultural Reform in the USSR', *OECD Observer*, No. 151, Paris, April/May.

Lewis, C. 1986, *The Role of Biotechnology in Assessing Future Land Use within Western Europe*, Commission of the European Communities, FAST Occasional Paper FOP 87, Brussels.

Lichtenberg, E. and D. Zilberman 1986, 'The Welfare Economics of Price Supports in US Agriculture', *American Economic Review*, December.

Lipton, K. 1989, *Changes in US Agriculture and Emerging Issues for Legislation in the 1990s*, USDA, Agriculture Information Bulletin No. 584, Washington DC.

Lukinov, I. 1989, 'Agrarian Policy and the Pricing Mechanism's Influence on the Agro-Industrial System', in EAAE (1989).

Mabbs-Zeno, C. and B. Krissoff 1989 *Tropical Beverages in the GATT*, Paper for the Joint OECD/World Bank Symposium on Agricultural Trade Liberalization: Implications for Developing Countries, OECD, Paris, 5–6 October 1989.

MacBean, A.I. and D.T. Nguyen 1987, *Commodity Policies: Problems and Prospects*, Croom Helm, London.

Marton, J. 1988, 'The Dynamics of the Hungarian Agricultural Structure', in FAO 1988.

Maskus, K. 1989, 'Large Costs and Small Benefits of the American Sugar Programme', *The World Economy*, 12, 1, March.

Miller, G., J. Rosenblatt and L. Hushak 1988, 'The Effects of Supply Shifts on Producers' Surplus', *American Journal of Agricultural Economics*, November.

Ministry of Agriculture, Fisheries and Food (MAFF) 1988, *Farm Incomes in the United Kingdom*, 1988 edition, London.

Molnar, J. and H. Kinnucan (eds) 1989, *Biotechnology and the New Agricultural Revolution*, Westview Press, Boulder, Colorado, for the American Association for the Advancement of Science.

Morozov, M. 1987, 'Selu Kak Gorodu' (to the Village as to the Town), *Kommunist*, Vol. 64, No. 1, January, Moscow.

Murphy, M. 1988, *Report on Farming in the Eastern Counties of England 1986/87*, Agricultural Economics Unit, Department of Land Economy, University of Cambridge, Cambridge.

Myers L., J. Blaylock and T. White 1987, 'Domestic and Export Demand for U.S. Agricultural Products', *American Journal of Agricultural Economics*, Vol. 69, No.2.

National Research Council 1989, *Alternative Agriculture*, US National Academy of Sciences, National Academy Press, Washington DC.

Neville-Rolfe, E. and C. Caspari 1987, *Potential for Change in the Use of Land in the European Community for Non-Food Purposes up to the Year 2000*, Commission of the European Communities, FAST Occasional Papers No. 178, Brussels.

OECD 1981, *The Instability of Agricultural Commodity Markets*, Paris.

OECD 1983, *Review of Agricultural Policies in OECD Member Countries 1980–1982*, Paris.

OECD 1987, *National Policies and Agricultural Trade*, Paris.

OECD 1988, *Agricultural Policies, Markets and Trade: Monitoring and Outlook 1988*, Paris.

OECD 1989a, *Biotechnology, Economic and Wider Impacts*, Paris.

OECD 1989b, *Agricultural and Environmental Policies: Opportunities for Integration*, Paris.

OECD 1989c, *Communiqué, Press Release*, Press/A(89)26, 1 June 1989.

Ollerenshaw, J. and F. Last 1988, 'Technology and the Environment', in Whitby, M. and Ollerenshaw, J. (eds), *Land Use and the European Environment*, Belhaven Press, London.

Osteen, C. and F. Kuchler 1986, *Potential Bans of Corn and Soybean Pesticides, Economic Implications for Farmers and Consumers*, USDA, Agricultural Economic Report 546, Washington DC.

Osteen, C. and L. Suguiyama 1988, *Losing Chlordimeform Use in Cotton Production, its Effects on the Economy and Pest Resistance*, USDA, Agricultural Economic Report 587, Washington DC.

Paarlberg, D. 1980, *Farm and Food Policy: Issues of the 1980s*, University of Nebraska Press.

Padoa-Schioppa, T. 1987, *Efficienza, Stabilità ed Equità: Una strategia per l'evoluzione del sistema economico della Comunità Europea*, Società Editrice Il Mulino, Bologna (Report of a Study Group appointed by the EC Commission).

Parikh, K., G. Fischer, K. Frohberg and O. Gulbrandsen 1988, *Toward Free Trade in Agriculture*, published for the International Institute of Applied Systems Analysis by Kluwer Academic Publishers, Dordrecht, The Netherlands.

Paulino, L. 1986, *Food in the Third World: Past trends and projections to 2000*, IFPRI Research Report No. 52, Washington, DC.

Penson, J. and B. Gardner 1988, 'Implications of the Macroeconomic Outlook for Agriculture', *American Journal of Agricultural Economics*, December.

Petit, M. 1985, *Determinants of Agricultural Policies in the United States and the European Community*, IFPRI Research Report, No. 51, Washington, DC.

Petit, M. 1988, 'The Agricultural Trade Confrontation between the United States and the European Community: A Challenge to our Profession', *Agricultural Economics*, 2.

Rausser, G., J. Chalfant, A. Love and K. Stamoulis 1986, 'Macroeconomic Linkages, Taxes and Subsidies in the US Agricultural Sector', *American Journal of Agricultural Economics*, May.

Research Institute of Agricultural Economics 1982, 'The Part of Hungarian Small-scale Farming in Production, Employment, Standard of Living and Way of Life', *Bulletin* No. 51, Budapest.

Robson, N., R. Gasson and B. Hill 1987, 'Part Time Farming: Implications for Farm Family Income', *Journal of Agricultural Economics*, Vol. 38, No. 2.

Roningen, V. and P. Dixit 1989, *Economic Implications of Agricultural Policy Reforms in Industrial Market Economies*, Economic Research Service, USDA, Washington DC, August 1989 (Staff Report No. AGES 89–36).

Ruttan, V. 1982, *Agricultural Research Policy*, University of Minnesota, Minneapolis.

Ruttan, V. 1987, 'Agricultural Research Policy and Development', *FAO Research and Technology Paper*, No. 2, Rome.

Sarma, J. and P. Yeung 1985, *Livestock Products in the Third World: Past Trends and Projections to 1990 and 2000*, IFPRI Research Report, No. 49, Washington DC.

Schilar, H. 1989, 'Planned Economy in the German Democratic Republic— Foundations and Changes' in UN/ECE, *Economic Reforms in the European Centrally Planned Economies*, Economic Studies, No. 1, United Nations, New York.

Schmidt, S. and W. Gardiner 1988, *Nongrain Feeds: EC trade and policy issues*, FAER-234, USDA, Washington DC.

Semenov, V. 1987, 'Soverchenstvovanie Finansovo Mekhanizma Agropromechlenovo Kompleksa' (Improvement of the Financial Mechanism of the Agroindustrial Complex), *Ekonomika Selskovo Khozyaistva*, September.

Shmelev, N. and V. Popov 1989, *The Turning Point, Revitalizing the Soviet Economy*, Doubleday, New York.

Stanton, B. 1986, *Production Costs for Cereals in the EC: Comparisons with the USA, 1977–1984*, Cornell University Agricultural Experiment Station (A.E. Res. 86-2), Ithaca, NY.

Tangermann, S., T. Josling and S. Pearson 1987, 'Multilateral Negotiations on Farm Support Levels', The World Economy, 10, 3.

Tikhonov, V. 1988, 'How to Spend the Billions', *Moscow News*, No.47, 1988.

Tirel J.C. 1987, 'Intensification Hier? Extensification Demain? Un Essai d'Analyse d'Images sur des Cliches Flous ...', INRA, Paris.

Tomic, D. 1989, 'Development and Problems in the Agriculture and Food-Processing Complex in Yugoslavia', in EAAE (1989).

Traill, B. 1988, *Technology and Food: Aims and findings of the EC FAST programme's research into the prospects and needs of the European food system*, Commission of the European Communities, FAST Occasional Paper 215, Brussels (forthcoming in the *British Food Journal*, 91 (1), Jan/Febr. 1989)

Treml, V. 1986, 'Soviet Foreign Trade in Foodstuffs', *Societ Economy*, Vol. 2, No. 1.

UNCTAD 1989, *Communication dated 8 March 1989 received from the Head of the Delegation of the USSR*, Document TD/B/1206, Geneva.

United Nations, Economic Commission for Europe (UN/ECE) 1987, *Economic Survey of Europe in 1986–1987*.

United Nations, Economic Commission for Europe (UN/ECE) 1988, *Economic Survey of Europe in 1987–1988*.

United Nations, Economic Commission for Europe and FAO (UN/ECE/ FAO) 1986, *European Timber Trends and Prospects to the Year 2000 and Beyond*, New York.

US Congress, Office of Technology Assessment 1986, *Technology, Public Policy, and the Changing Structure of American Agriculture*, Washington, DC.

US Council on Environmental Quality and Department of State 1980, *The Global 2000 Report to the President*, US Government Printing Office, Washington, DC.

USDA 1983, *Economic Indicators of the Farm Sector, Income and Balance Sheet Data, 1983*, ECIFS 3-3, Washington, DC.

USDA 1986a, *Economic Indicators of the Farm Sector: National Financial Summary, 1985*, ECIFS 5-2, Washington, DC.

USDA 1986b, *Fuel Ethanol and Agriculture: An economic assessment*, Office of Energy, Agricultural Economic Report No. 562, Washington, DC.

USDA 1987a, *Economic Indicators of the Farm Sector, Farm Sector Review, 1985*, ECIFS 5-4, Washington, DC.

USDA 1987b, *1987 Fact Book of U.S. Agriculture*, Miscellaneous Publication No. 1063, Washington, DC.

USDA 1987c, *Year-End Update, U.S. Agriculture*, Washington, DC.

USDA 1988a, *Agricultural Resources: Cropland, water and conservation, situation and outlook report*, AR-12, Washington DC.

USDA 1988b, *Ethanol: Economic and policy tradeoffs*, Washington, DC.

USDA 1989, *World Agriculture: Situation and outlook report, special issue, 'Are we approaching world food crisis again?'*, WAS-55, Washington, DC.

Valdés, A. and J. Zietz 1980, *Agricultural Protection in OECD Countries: Its cost to less developed countries*, IFPRI Research Report, No. 21, Washington DC.

Willis, K., J. Benson and C. Saunders 1988, 'The Impact of Agricultural Policy on the Costs of Nature Conservation', *Land Economics*, Vol. 64, No. 2.

Winters, A. 1987, 'The Economic Consequences of Agricultural Support: A Survey', *OECD Economic Studies*, No. 9, Autumn.

World Bank 1986, *Price Prospects for Major Primary Commodities*, Document 814/86, Washington DC.

World Bank 1989, *Price Prospects for Major Primary Commodities 1988–2000*, Washington DC.

Wos, A. 1988 'The Dynamics of Agrarian Structure in Poland', in FAO (1988).

Young, D. 1988, 'Economic Adjustment to Sustainable Agriculture: Discussion', *American Journal of Agricultural Economics*, December.

Zietz, J. and A. Valdés 1989, *International Interactions in Food and Agricultural Policies: Effect of alternative policies*, Paper for the Joint OECD/World Bank Symposium on Agricultural Trade Liberalization: Implications for Developing Countries, OECD, Paris, 5–6 October 1989.

Index